LECTURES ON GEOLOGY

DR. JOHN WALKER OF EDINBURGH

From "Memoir of John Walker, D.D." in Sir William Jardine's *Natural History of the Birds of Great Britain and Ireland* (Edinburgh, 1842).

LECTURES ON
GEOLOGY

INCLUDING HYDROGRAPHY, MINERALOGY, AND METEOROLOGY
WITH AN INTRODUCTION TO BIOLOGY

BY

John Walker

PROFESSOR OF NATURAL HISTORY AND
KEEPER OF THE MUSEUM, UNIVERSITY OF EDINBURGH
FROM 1779 TO 1803

EDITED WITH NOTES AND AN INTRODUCTION

BY

Harold W. Scott

THE UNIVERSITY OF CHICAGO PRESS · CHICAGO & LONDON

Library of Congress Catalog Card Number: 65-24986

The University of Chicago Press, Chicago & London
The University of Toronto Press, Toronto 5, Canada

To my wife, Joann
whose enthusiasm for this study has been continuous

PREFACE

John Playfair, Sir James Hall, and Robert Jameson occupy important places in the list of pioneers of geology. The geological profession always quotes the contributions made by these men in any history of the science. The praise bestowed upon them is indeed well deserved, but it is sad to contemplate the fact that John Walker, their teacher, has been completely neglected. My object in compiling this volume of Walker's lectures on geology has been to show the nature of the teaching such men as Playfair, Hall, and Jameson received at his hands. The lecture notes also, it seems to me, establish Walker as one of the first great figures in the history of geology.

My interest in John Walker dates from a visit to Edinburgh in 1960. In the course of other work at the university I came upon references to Walker indicating that he might be a figure of some interest in the history of geology. Mr. Charles P. Finlayson, Keeper of the Rare Manuscript Library, introduced me to the voluminous Walker papers owned by the university library. After reading through them all—and becoming convinced myself of Walker's importance—I resolved to make a selection of his geological work available to others with an interest in our field. The selection printed in this volume comprises all Walker's surviving lectures on natural history except the series on the "Vegetable and Animal Kingdoms." None of this material has been previously published, unless Walker's printing of his mineral classification for classroom use may be deemed publication. I have also included, as of general interest, translations of some surviving correspondence between Walker and Linnaeus; this too has not been fully published until now.

THE MANUSCRIPTS

The manuscripts of Walker's lectures constitute several volumes, as may be seen from the Bibliography. Some are copies of the lectures made by students (including Jameson, whose copy is almost illegible); others seem to have been prepared by a professional copyist. The copies that can be dated run from 1782 to some time after 1796 (watermark date on the paper). All copies are essentially similar: they cover the same subject matter and often use the same phraseology; they differ, however, by occasional omissions or additions. The section on "Bowlder Stones," for example, is lacking in the 1782 copy but appears in later copies. Walker kept his notes up to date, and many of the revisions in the lectures are additions representing his analysis of new literature. Thus we find materials dated 1788 and 1789 covering discoveries made during those very years.

The lectures printed in the present volume depend on several of the manuscripts, but most heavily on the one designated D.C. 10-33. It is the most complete and is among those most readily transcribed. A note on the first page, "c–1788," apparently means that it was copied about that date but gives no hint of the date of original composition. No doubt is is, like the other manuscript versions of the lectures, essentially what Walker set down between 1779 and 1781 when he initiated the course in natural history.

Transcription of the lectures was beset by many editorial problems. The chief difficulty arose from the fact that the manuscripts were not in Walker's handwriting. Mr. Finlayson says that D.C. 2-18 and D.C. 2-24, for example, "are from sets that appear to be in the hands of scribes, perhaps written for sale, copied from a master-copy possibly compiled in some sort of short hand." Names in the manuscripts often appear to be phonetically spelled; for instance, Rammel for Ramel, Aminton for Amontone, M. Laurin for Maclurin. Several names are omitted entirely, a blank space being left in the line at that point. Capitalization and punctuation are inconsistent and sometimes absurd. None of these blemishes, it should be noted, characterize Walker's own writing, as illustrated in his letters to Linnaeus or in his published articles and books.

The chief question has thus been: Should the manuscripts be reproduced as found, or should changes be made in capitalization, spelling, and punctuation in accord with Walker's usual style? It has seemed more important to me to present Walker's work in a form reasonably easy to follow than to produce a facsimile copy of the text, including scribal errors. Therefore

I decided to make a few corrections in the text and to interpret symbols, spelling, and punctuation in certain consistent ways. These may be summarized as follows.

Spelling. Most variant spellings common in the eighteenth century (alledge, shew, smoak, and so on) have been left as found, even though this involved some inconsistency. Such forms as "diarys" have also been left. Common nouns and adjectives capitalized in the manuscript have been left, but where proper nouns are spelled with lowercase letters I have silently capitalized them. The apostrophe is used inconsistently in the manuscripts (often to mark the plural of a noun, for instance); I have left these. Numerals, also, are inconsistently used—sometimes arabic, sometimes roman—and I have not changed them. Accents have been left as found on French words (abbè), but I have added umlauts to German words where needed.

The following words and proper names have been silently regularized: balance, character, deficiency, embouchure, Fahrenheit, Hooke, inches, inflammable, Isaac, isthmus, Mississippi, Musschenbroek, ordnance, predilection, Priestley, putrefaction, rarefaction, Saussure, siliceous, and venomous.

Abbreviations have been regularized to some degree. In all cases I have lowered the superior letter and added a period to such abbreviations as D^r, M^r, and the like, and have spelled out in full such abbreviations as B^p (Bishop), Sp^t (Spirit), and w^{ch} (which). I have also spelled out & and converted &c to etc. Finally, I have added periods to abbreviations for the compass directions, N, E, S, W.

Punctuation. It is in punctuation that the manuscripts depart most markedly from modern or normal eighteenth-century practice. In addition to the period, a colon is often used for a full stop, as is (sometimes) a wide space in the line and (rarely) a flourish of the pen. Less than a full stop was represented by a semicolon (often difficult to distinguish from a colon), a comma, or again, simply a wide space. Punctuation at the ends of lines and paragraphs was frequently omitted, although some omissions are probably due to fading of the ink. I have attempted to represent full stops consistently by periods and to supply appropriate punctuation wherever the manuscript uses wide spacing. In the latter cases, where capital letters were lacking at the beginning of sentences (as they frequently were), I have often had to decide simply on the basis of sense whether a certain stop should be a period, a semicolon, or a comma. Sometimes no amount of punctuating would give a construction fitting modern rules of syntax: if such a sentence made sense, I simply left it. Also left as found were all parentheses.

Identification of persons mentioned. Walker referred to a great many authorities in his lectures, using last names only. Copyists have misspelled some names and omitted others, leaving a blank in the manuscript where the name ought to be. I have tried to identify all persons referred to (except well-known names like Aristotle or Newton), but have not always been successful. I have indicated a blank left for a name by ———— or, where the context permits a conjecture, by a name and question mark in brackets. I have attempted to indicate for each man his area of interest, his nationality, or the dates of his birth and death. In most instances I have also tried to give the name of his best-known work or the work Walker apparently had in mind in referring to him.

Miscellaneous. Paragraphing in the originals is usually clear; I have made no changes, even though it is often inconsistent and not in accord with modern principles. I have used brackets to enclose anything I have added, except for the silent corrections mentioned above. Additions may be a word or letter, a question mark (to indicate a questionable interpretation), or the word *sic* (to flag an unusual error in spelling or grammar).

I hope that in this editing I have done justice both to Walker and to the reader.

ACKNOWLEDGMENTS

The book has been made possible by the co-operation of many friends. The American Council of Learned Societies awarded me a grant which made it possible to revisit Edinburgh during the summer of 1962. This assistance is gratefully acknowledged.

Special thanks are due to Mr. Charles P. Finlayson, Keeper of the Rare Manuscript Library at the University of Edinburgh. Without his aid this material could never have been assembled. His patience, good humor, and knowledge made my work very pleasant.

Dr. George White, professor of geology, University of Illinois, has shown a constant enthusiasm for this investigation; his knowledge of Playfair and Hutton, his reading of a part of the manuscript, and his many suggestions have been most helpful. Dr. Albert Carozzi's knowledge of Werner's works has been used to great advantage. I am indebted to the Rev. Dr. W. M. Laing, former minister of St. Cuthbert's, the Colinton parish church, for his gracious hospitality and for information about Walker; and to his successor, the Rev. W. B. Johnston, for numerous favors since my return.

To all my colleagues in the Department of Geology, I express my thanks for help with specific problems. To Professor G. Neville Jones, Department of Botany, University of Illinois, I am indebted for help in clarifying several botanical questions. Professor John Heller, Department of Classics, assisted in interpreting the Greek and some Latin. Dr. Robert Hansman, Department of Geology, and Mr. Frank Rodgers, University of Illinois Library, have assisted with editorial work, and I am pleased to acknowledge their constructive help.

To the Department of Botany, University of Edinburgh, I am indebted for the copy of Walker's portrait, and to the staff of the geology section of the Royal Scottish Museum I owe special thanks for their courtesy and aid.

I am especially indebted to a friend who has helped financially to make the publication of this book possible.

CONTENTS

ILLUSTRATIONS

BIOGRAPHICAL INTRODUCTION

Walker as a Geologist

"The objects of nature themselves must be sedulously examined in their native state, the fields and the mountains must be traversed, the woods and the waters must be explored, the ocean must be fathomed and its shores scrutinized by everyone that would become proficient in natural knowledge." With these ringing words the Rev. Dr. John Walker admonished his students at the University of Edinburgh, as early as 1781. The way to knowledge of natural history is to go to the fields, the mountains, the oceans, and to observe, collect, identify, experiment, and study.

John Walker was appointed regius professor of natural history at the University of Edinburgh in 1779. This was several years in advance of what is commonly thought of as the beginning of modern geology. Geology had not yet emerged from the Dark Ages. Witchcraft and fantasy often still marked discussion of volcanoes, earthquakes, disappearing springs, and fossils. Physics had been long emergent, chemistry was well on its way, botany and zoology had attained some degree of respectability, but geology did not exist; only mineralogy as a separate discipline had any standing. It fell to the lot of John Walker to raise the banner of geology at the University of Edinburgh and carry it for twenty-four years. He was destined to become a forgotten man, whereas his students Playfair, Jameson, and Hall, and his brilliant contemporary Hutton, would receive the accolades of future generations.

From Aristotle to Walker the treatment of geological processes was for the most part a succession of hypotheses built upon nothing more substantial than imagination and a few disconnected observations; fact and fancy still were confused. Most of the other sciences had broken with the past, but geology was yet to be born. As we read Walker's lectures we become aware that even though some of his concepts were false, he represented the beginning of a new age in geology. He is the connecting link between medieval and modern geology; he represents the beginnings of the new and the discarding of the old. It was Walker's role to give life to a new science, organizing and creating lectures in a discipline never before taught as such.

John Walker was born in Edinburgh in 1731. His father, rector of Canongate Grammar School, gave him early training in Latin, and we are told that at the age of ten the boy enjoyed reading Homer—presumably in Greek. His scientific interests began early, perhaps aroused by reading (also at the age of ten) Sutherland's *Hortus Edinburgensis*. Botany was not his only interest, however. "I began to collect minerals in the year 1746," he later wrote. For many years, these two interests were to share his attention.

After finishing studies at his father's school, he entered the University of Edinburgh to prepare for the ministry in the Established (Presbyterian) Church of Scotland. At the university he was greatly attracted by the chemistry course offered by Professor William Cullen. Cullen was interested in minerals, and the enthusiasm of young Walker for mineral study quickly caught his attention. "I attended his course of chemistry two winters," Walker said, "and, being favoured with his friendship and intimacy, I became more and more attached to mineralogy." With Cullen, with other friends, or alone he made a number of collecting trips while at the university.

Another attraction of the university for the young Walker was the museum display of natural history. At that time the museum, which represented the combined collections of Sir Andrew Balfour and Sir Robert Sibbald, was of considerable importance. Walker referred to it with great respect: "Both together formed a Museum as respectable as any University in Europe could at that time boast of. . . . I may well remember it for it was the view of it which first inspired me with a taste for Natural History when a boy at this College. . . . By inexcusable neglect this valuable collection perished."

Licensed to preach by the presbytery of Kirkcudbright in 1754, Walker was ordained minister and called to his first post, at Glencorse, in September, 1758. That same autumn he wrote his first scientific paper, an account of a medicinal well. At about the same time, even more significantly, he made the acquaintance of Henry Home, Lord Kames, one of the most prominent Scottish leaders of the time. Home possessed energy and intelligence and had an abiding interest in the welfare of his country and its people. He also had a great thirst for knowledge of the earth and its inhabitants. Finding in Walker a kindred spirit, he soon came to rely on him for all sorts of information. The acquaintance, which developed into a relationship of mutual respect and admiration, changed the entire course of Walker's life. It was this connection that gave the young minister the opportunity to make his contributions to science and education.

When transferred to Moffat in June, 1762, Walker immediately began to search the fields and mountains, breaking rocks, gathering minerals, visiting mines, collecting flowers and "even weeds." These habits seemed mysterious to the local inhabitants, and he became known as the "mad minister from Moffat." Continuing the habits he formed as a boy, Walker traveled far in pursuit of mineral specimens for his collection, which was becoming extensive. Among other places, he journeyed to

the lead-mines at Mackrymore, the copper mines at Covend, and the mines of antimony at Eskdale. . . . I may have been at these mines about thirty times. . . . Between the years 1761 and 1764, I found in those mines the *Strontianite,* the *Ore,* and the *Ochre of Nickel;* the *Plumbum pellucidum* of Linnæus; the *Plumbum decahedrum* and *cyaneum,* both undescribed; the *Saxum metalliferum* of the Germans; the *Ponderosa aerata* of Bergman; and the *morettum,* which afterwards appeared to be a peculiar sort of Zeolite. All these were here, for the first time discovered in Britain. . . . I found several rare minerals, and particularly that singular substance, since called Strontianite, in great plenty; though I had observed it sparingly, three years before, in the Mines of Leadhills.[1]

It is of interest that Walker collected strontianite almost thirty years before it was formally named and described by T. C. Hope,[2] one of his students.

THE HEBRIDES SURVEY

After two years at Moffat, Walker was commissioned, through the influence of Lord Kames, to make a survey of the Hebrides and the Highlands. The Hebrides were poor and economically undeveloped and their inhab-

[1] John Walker, "Notice of Mineralogical Journeys, and of a Mineralogical System, by the late Rev. Dr. John Walker," *Edinburgh Phil. J.* 6 (1822): 88–95. In a footnote Jameson comments: "It is not generally known, that at one period, small quantities of strontites were found at Lead Hills; and the fact in the text proves, that to Dr. Walker the merit is due of having determined mineralogically that Strontites was a new mineral species. Dr. Hope afterwards, by the discovery of the strontitic earth, added to the interest of the determination of Dr. Walker, and proved that strontites was also a new chemical species."

[2] T. C. Hope, "Account of a Mineral from Strontian . . . ," *Trans. Roy. Soc. Edinburgh* 4, Pt. 2 (1798): 3–39. Read Nov. 4, 1793.

itants underprivileged. Lord Kames wished to improve the situation, and Walker was chosen for the job of making a survey. "A man most eminently qualified for that employment, as joining to every endowment of scientific knowledge, requisite for the undertaking, an ardent mind, and a great portion of natural sagacity and penetration"—thus Home described Walker. At the same time Walker was asked to make a report to the Society for the Propagation of Christian Knowledge.

The trip, which lasted from May to December, 1764, was not Walker's first to the Hebrides and Highlands, nor would it be his last. He studied and traveled extensively in those areas from 1760 to 1785. After the first commission in 1764, he reported to the General Assembly in 1765; recommissioned in 1771, he again reported in 1772. Later he said, "The report to the Annexed Board formed a large volume, which remained for some time in possession of the Board, but was afterwards sent to London, and of which no exact copy was retained. This volume has since disappeared, and even after much enquiry to recover it, has been given up as lost."

The Hebrides report was first published in 1808 by Walker's friend Charles Stewart, who found the manuscript in Walker's library after his death.[3] Stewart seems to have collected various articles that Walker wrote on the Hebrides and assembled them in two volumes for printing. Most of the observations were certainly made in 1764 and 1771, but some data, such as population statistics, were not compiled until 1795. As in so much of Walker's work, the date of a specific observation is difficult to establish.

The first Hebrides survey was carried out under the most trying circumstances. Walker noted in a letter to Lord Kames that the work required more than 3,000 miles of travel by foot, horse, and boat. He visited all the islands, investigating the mineral and plant resources as well as the agricultural and fishing industries. He took a census of the people and studied their habits and economy. Hardly any item of importance escaped his notice.

The survey was more than a casual review of the situation. It was an enormous one-man task and in many respects included what a natural history survey of today might cover with a team of experts. As later published, the report occupied 805 printed pages. The subject treated most thoroughly was agriculture, but the report included sections on geology, botany, zoology, industry, and commerce, as well as remarks on the character and habits of the people.

Walker's treatment of minerals in the Hebrides report is particularly interesting. He mentions seven kinds of marble, two kinds of granite, two kinds of serpentine, porphyry, jasper, onyx, chalcedony, soap rock, tripoli,

[3] John Walker, *An Economical History of the Hebrides.*

conglomerate, "chrystalline" sand, calcareous stone, and micaceous rocks. Of the granites he calls one syenite or red granite. "The high mountain of Cruschin, in Argyllshire, consists of syenite or red granite, being the same stone with that at Pompey's pillar at Alexandria."[4] And of gray granite or Moor stone, "the mountain of Crufelt in Galloway is composed." Among the minerals and rocks that he found, "the beautiful carnation marble of Tirey; the white marble of the same island, with green transparent schorl; the white statuary of Skye; the green serpentine and *Lapis nephriticus* of Iona; the obsidian of Eig; the green jasper of Rume; the aminatine rock of Bernerey; and the black lead of Glenelg, were then first made known."[5]

Later he told how his old friend Professor Cullen was the instigator of one discovery. "Not long before I set out Dr. Cullen had received the first German edition of Cronstedt's Essay, of which he was so fond, that he carried it for several weeks in his pocket. . . . He was particularly anxious about the Zeolite; and it was in consequence of this, that I first observed it among the basaltic rocks at the Giants Causeway."[6]

While at Moffat, Walker turned his most serious attention to the study of botany. The general flora of Scotland, especially the plants responsible for the production of peat, were of interest to him. Problems of reproduction, particularly in the mosses, held the greatest attraction for him. It was during this period too that he began to correspond with Linnaeus, and we find him writing his first letter to the great botanist on January 8, 1762. Walker's expression to Linnaeus of his strong interest in botany should not be taken to mean that he was limiting his studies to plants. In fact, Walker was never capable of restricting himself to one subject. Many of his *Essays,* papers, mineralogical journals, and his preliminary mineral classification were written during this period.

From the Hebrides Walker returned to Moffat in 1765 to submit his report and to resume his ministerial duties. Soon after his return he was granted degrees by two Scottish universities: a doctorate of divinity by Edinburgh and an honorary doctorate of medicine by Glasgow. The year marked a turning point for Walker. In a sense the die had been cast: henceforth he would

[4] *Ibid.*, p. 384. Ben Cruachan, a granite mountain in southwest Scotland, is 3,680 feet in elevation. Ben Crufell (1,825 feet) is in northwest Dumfriesshire, six miles southwest of Sanquhar.

[5] Walker, "Notice of Mineralogical Journeys," p. 90.

[6] Axel F. Cronstedt, *Försök til Mineralogie, eller Mineral-rikets Upstallning* (1758). Walker refers here to the early, little-known German translation of Cronstedt: G. Wiedemann, *Versuch einer neuen Mineralogie aus dem Schwedischen übersetzt* (Copenhagen, 1760). This explains the source of Walker's early knowledge of Cronstedt. He did not refer to Werner's later translation of 1780.

divide his energies between religion and science. He would conduct the services on Sundays and perform his parish duties, but much of his thought and time would be given to the study of rocks and minerals, plants, and animals; to exploring mines and caves, collecting, classifying, reading, traveling, and writing.

THE UNIVERSITY APPOINTMENT

The Hebrides report revealed Walker as a man of unusual ability, with keen powers of observation, high intellectual capacity, and great physical vigor. These assets were not lost upon Lord Kames, and in 1779, when the chair of natural history at the University of Edinburgh became available upon the death of Robert Ramsay, Walker received the appointment, and during the same year he was also appointed keeper of the museum. There was probably no man in the British Isles better qualified for these positions.

Although Ramsay had been the first professor to occupy the chair, Walker was the first to become active. There is no evidence that Ramsay ever gave lectures, held class meetings, or prepared syllabi for students. Formal geological education had its beginning in Walker's classroom. Walker was prepared for his new duties by temperament, education, extensive field experience and personal research, a thorough knowledge of all the existing scientific literature, and an acquaintance—at least by correspondence—with some of the leading investigators of the day. He was forty-eight years old, mature in years as well as scientific experience. His stipend of £68 a year as professor was not much encouragement. This was supplemented by student fees of £3 5s.—but records show that some could not pay. But the records show also that students did come, many from America as well as Great Britain, Jamaica, South America, and Europe. The 1793 class, for example, was composed of forty-eight students, including three from London, two from Ireland, one each from Jamaica, Bahama, Antigua, Rio de Janeiro, and South Carolina. Most, including many of the Americans, were students of medicine.[7]

WALKER *and* HUTTON

There is not much in the record concerning the relationship between Walker and James Hutton. Future work may uncover professional relationships not now known. How much influence each man had upon the other is

[7] Cf. "The Indiana Physician as Geologist and Naturalist," *Magazine of Indiana History* 56, No. 1 (1961):1–35. Physicians were among the early naturalists of America and made contributions to various sciences. No fewer than eight physicians contributed to the geology of Indiana. It would be of interest to know what happened to the Walker-trained American physicians.

difficult to say. Walker's work, however, was certainly done independently of Hutton. They were both members of the Philosophical Society, and upon the formation of the Royal Society of Edinburgh both became members. Thus it appears that the two men must have been acquainted for many years. We can only conclude they knew each other, were on friendly terms, but did not actively exchange information except in formal meetings at the Royal Society of Edinburgh or at informal meetings of a semisocial nature, such as the meetings held by Lord Kames, in the form of levees or suppers attended by learned men of the Edinburgh community. Most of Walker's field work was done and manuscripts written between 1758 and 1781 and therefore are older than Hutton's dissertation read in 1785 and his two volumes entitled *Theory of the Earth* published in 1795.[8]

WALKER *and* PLAYFAIR

John Playfair, one of Walker's adult students and also a minister, was later to become famous as the interpreter (1802) of Hutton.[9] Playfair was appointed professor of mathematics in the University of Edinburgh in 1785, and to the chair of natural philosophy in 1805. After earning his M.A. degree at the University of St. Andrews in 1765, Playfair became a parish minister. He spent some time in Edinburgh in 1769, where, among other intellectuals, he met James Hutton. In 1782 Playfair moved to Edinburgh as a private teacher for the sons of the Ferguson family. At this time, at the age of thirty-four, he enrolled in Walker's class. His interest in geology may well have been aroused by Hutton before his enrollment, but part of his geological education was certainly acquired from Walker. Walker and Playfair attended meetings of the Senatus Academicus as professors of equal rank. Also, after the retirement of James Gregory as joint secretary with Walker of the Physical Section of the Royal Society of Edinburgh, Playfair was elected to the vacancy, and Walker and Playfair acted as joint secretaries for several years.

[8] James Hutton, "Concerning the Systems of the Earth, Its Duration, and Stability" (abstract of a dissertation read before the Royal Society of Edinburgh, 1785). See V. A. Eyles, "A Bibliographical Note on the Earliest Printed Version of Hutton's Theory of the Earth," *J. Soc. Bibl. Nat. Hist.* 3, Pt. 2 (1955); Hutton, *Theory of the Earth with Proofs and Illustrations* (2 vols.; Edinburgh, 1795).

[9] John Playfair, *Illustrations of the Huttonian Theory* (Edinburgh, 1802); reprinted with an introduction by George W. White (Urbana: University of Illinois Press, 1956). One of Playfair's classmates was Sir James Hall. After spending two years (1777–79) at Christ's College, Cambridge, Hall studied in France and Geneva, returning to Scotland in 1782. It is at this time that we find his name along with that of Playfair listed in Walker's class.

WALKER *and* HALL

The name of Sir James Hall of Dunglass, sometimes referred to as the founder of experimental geology, appears at the head of Walker's class records in the November section of 1782. This was the same class attended by John Playfair. After a short period of study at Cambridge, Hall visited Europe, and upon his return enrolled in Walker's class. In 1788, Hall accompanied Playfair and Hutton to the Lammermuir Hills, where Hutton pointed out an unconformity and interpreted its meaning. This meeting in the field was later described by both Hall and Playfair and has become one of the famous moments in the history of geology.

WALKER *and* JAMESON

Another student, Robert Jameson, was to become Walker's protégé and gain fame as a mineralogist and pioneer geologist, as well as a disciple of Abraham Gottlob Werner. It is not clear just when Jameson first knew of Werner. He referred to Werner several times (1798 and 1800) and stated his belief in Werner's concepts of the origin of granite. No mention of Werner has been found in Walker's manuscripts, nor has any comment been found in Walker's notes concerning Werner's concepts. Walker apparently did not discuss the origin of granite. Jameson's ideas on this matter appear to have been wholly derived from Werner.

From Jameson's class days to 1800 he was in most respects serving an apprenticeship under Walker; he made trips, collected and prepared material, and ran errands. About 1792 Walker turned much of his museum work over to his protégé. In 1800 Jameson left for a period of study in Freiburg with Werner which was to last until April, 1802, when he returned to Edinburgh.

Jameson acknowledged his debt to his old teacher in many ways: first, by dedicating two books to him; second, by including Walker's classification of minerals along with those of Cronstedt (1758), Werner (1789), Kirwan (1794), and others;[10] third, by many references to his works; fourth, by adopting in his mineralogical reports the *Mineralogical Journal* methods which Walker had used (*ca.* 1770); and finally by publishing in 1822 Walker's paper, "Mineralogical Journeys and a Mineralogical System."

The Walker-Jameson relationship was certainly a close and warm association, marked by the mutual respect of a grateful student and an honored professor. The relationship was so close that Jameson often found himself repeat-

[10] Robert Jameson, *A System of Mineralogy*, Vol. I.

ing not only the substance but even the details of Walker's teaching in almost the same words. Jameson's early writings are filled with "Dr. Walker states . . . Dr. Walker believes . . . Dr. Walker finds. . . ." There is no question of Jameson's acknowledgment of his debt to Walker.

Jameson's work on the Shetland Islands in 1798 and his *Mineralogy of the Scottish Isles* in 1800[11] were both prepared before his excursion to Freiburg and during and after his training under Walker. His 1800 work is filled with references to Werner, and it is obvious that he had adopted by this date the main concepts of Werner concerning the origin of granite and basalt. His reports, for the most part, were based on observations made in the field, and followed the broad general principles and methods he had been taught, but Walker did not teach him that basalt was not volcanic or that granite was precipitated from water. Upon his return from Germany and his appointment as regius professor of natural history, Jameson introduced the doctrines of Wernerism into the classroom and also formed the Wernerian Society.

The great biographer Geikie[12] knew that Jameson had studied under Werner, but it is surprising that he did not realize that Jameson already had a geological education and had written two substantial books on geology before going to Freiburg. Geikie, referring to the Edinburgh group, says, "Among these men there was only one teacher—the gentle and eloquent Playfair." Jameson did not suddenly appear in Freiburg without a background in geology any more than Playfair or Sir James Hall suddenly saw the new geological frontier. The frontier had been pointed out to them many years before in Walker's classroom and laboratory. This is not to say that Hutton had no influence on Playfair; he did, and at an early date. Ultimately, Jameson followed Werner, whereas Playfair followed Hutton, and Walker's contributions to the educational development of Hall, Jameson, and Playfair were lost.

KEEPER *of the* MUSEUM

Unfortunately, Walker's second position as keeper of the museum carried no stipend and he had no working budget; yet he was genuinely interested in the museum. He recognized its value in teaching and also its importance to the public. The extensive collections of Balfour and Sibbald had been emasculated before Walker's arrival, depleted beyond hope of recovery. Walker's own collections were fairly substantial, but he needed and wanted much more.

[11] Jameson, *Outline of the Mineralogy of the Scottish Isles,* Vols. I and II.

[12] Sir Archibald Geikie, *The Founders of Geology* (London: Macmillan & Co., Ltd., 1897), p. 325.

DEDICATION PAGE BY ROBERT JAMESON

A real affection for his old teacher (by then blind and retired), as well as an intellectual debt,
seems to underlie Jameson's dedication to Walker
of his *Mineralogy of the Scottish Isles.*

By order of the Edinburgh town council[13] he was ordered to prepare "a full list or inventory of all Curiousities or Rarities belonging to said University." This list was received by the council on March 22, 1780.[14]

The charter of the Royal Society of Edinburgh ordered that "all objects of natural history presented to the Society shall be deposited in the Museum of the University of Edinburgh."[15] Many specimens were received by Walker for the museum and most of the gifts were listed in the *Transactions*. These gifts Walker acknowledged and catalogued, but, unfortunately, after his death the material disappeared.

Without funds Walker could make no progress at the museum. Acquiring subscriptions for purchases was difficult and slow, and the substantial financial support needed for a major museum never materialized. It is no wonder that we find a note of irritation in the letter written in 1785 to the town council requesting funds: "'I have now for six years, been arduously employed in establishing a public museum in the College. I have a commission from the King, and another from the City of Edinburgh to be Keeper of the Museum; but when I settled here there was nothing to keep."[16]

In his lectures he states that after materials are collected and prepared, they must be properly placed in a museum. But "for our own country (I am sorry to say it) it is farther behind in this respect than any other in Europe except those of Portugal and Spain."[17] After a further discussion of museums and after expressing his regrets at being unable to acquire funds, he states, "I must therefore content myself with having recourse to my own little private collection which tho considerable enough for my opportunities, yet is altogether inadequate."

Despite these words, Walker's private collection was far from inadequate. For his mineral and rock collection he received material from many parts of the world through friends; these included a specimen of Niagaran dolomite from Niagara Falls, geyserite from Iceland, minerals that Pallas sent him from Russia, fossil ferns from Pennsylvania, sulfur from the West Indies, and minerals from Newfoundland. It is not clear how Walker obtained the American specimens. It is possible that some came from Benjamin Franklin —Walker is known to have conferred with Franklin upon the latter's visit to

[13] Council Minutes (City of Edinburgh, Nov. 3, 1779), Vol. 98.

[14] *Ibid.*, p. 316.

[15] *Trans. Roy. Soc. Edinburgh* 1 (1783):13.

[16] Letter in Rare Manuscript Library, University of Edinburgh.

[17] See text, p. 46.

Edinburgh[18] but it is more likely that some of his many American students acquired the samples for him. His laboratory was equipped with more than 3,000 specimens, representing over 1,500 species of rocks and minerals. This was indeed a substantial collection for teaching and some modern laboratories are not equally well supplied. It was probably exceeded by none in the latter half of the eighteenth century with the possible exception of Werner's.

Upon Walker's death the trustees apparently removed most of his collection from the museum, although a remnant of it remains in the present museum, consisting mostly of marbles classified as various species and varieties under the genus *Marmor*. Numbers run from 210 to 388 and include, in addition to *Marmor*, *Phengites* a travertine, *Leucosticites* a porphyry, *Saurites* a marble, *Syenite* an acid-intrusive approaching granite, and *Catochites* a basic-intrusive. An 1822 footnote by Jameson[19] suggested that the collection was going to be sold. Where or in what manner it disappeared is unknown. How or why Jameson allowed it to get out of museum control is equally a mystery.

Society Member

In addition to performing his duties as keeper of the museum Walker helped organize the Royal Society of Edinburgh. His name is given in the charter of the Society,[20] and he was an officer and active member throughout the remainder of his life. He was appointed first secretary of the Physical Section, jointly with James Gregory in 1783, and later jointly with John Playfair until 1796, thus serving in this capacity for thirteen years. The first article printed in the *Proceedings* of the society, Vol. 1 (1788), was by Walker, the second by Playfair, and the third by Hutton. This was indeed a triumvirate of talent.

The Natural History Society of Edinburgh was formed in 1782 and Walker became a member the same year. Several of his unpublished manuscripts are preserved as handwritten copies in the records of this society.

Walker carried on an interesting service for anyone who wrote him for information. Acting somewhat as a "one-man survey," he gave advice on rocks and minerals, plants, and agriculture in general. His "how to do it" or "what is it" service was basically the same as that performed by natural history and geology surveys instituted more than a half-century later by governmental agencies. His vision of service was so far ahead of his time that he

[18] John Walker, *Essays on Natural History and Rural Economy;* "A catalogue of some of the most considerable trees in Scotland," p. 2.

[19] Walker, "Notice of Mineralogical Journeys," p. 94.

[20] *Trans. Roy. Soc. Edinburgh* 1 (1783):7.

had no chance of getting major financial support. His work on the Hebrides in 1764 almost certainly showed him the value of mineral and agricultural reports to the welfare of the general public.

WALKER *as an* AGRICULTURIST

In Scotland the Highland Society had given impetus to agricultural studies as early as 1784 with John Walker as one of the leaders. They made strenuous efforts to find means of increasing production on the thin soils of Scotland, especially in the Highlands and Hebrides. Walker had studied this matter since 1764. Out of his interest in agriculture and his work with the Highland Society he gave a series of lectures on agriculture in 1790.

Lectures on Agriculture

On Thursday last, Dr. Walker, Professor of Natural History in the University here, concluded the first course of Lectures on Agriculture, which has ever been delivered in Britain as a branch of Academical education. The gentlemen who attended that class invited him afterwards to an entertainment, that they might have an opportunity of expressing to him collectively their acknowledgments and thanks for the instruction they had received: and at that meeting an Agricultural Society was projected, which, under his patronage and direction, may prove of essential service to the practical farmer, and tend to the general diffusion of Georgical science over the country.[21]

Walker organized the Agricultural Society of Edinburgh and was an unsuccessful candidate for the position of professor of agriculture at Edinburgh University in 1792.[22]

Shortly before 1800 a severe food shortage developed in Scotland. Walker published a paper on the subject in which he correctly evaluated the cause of the shortage as due to the per acre productivity of meat as compared to grain and recommended a return to a grain economy.

No attempt has been made to evaluate his work in the field of agriculture, but a preliminary examination makes it appear his contributions may have been considerable. Certainly, he was among the earliest men to give lectures on the university level, preceded only by Hungarian institutes. He had a position of leadership in Scotland in the late eighteenth century and was the first English-speaking man to introduce agriculture into higher education.

[21] *The Caledonian Mercury* (Edinburgh, April 3, 1790).

[22] Andrew Coventry, M.D., took office as the first professor of agriculture at the University of Edinburgh on Nov. 17, 1790, and held this position until 1831.

INVESTIGATOR *and* TEACHER

Walker was no armchair theorist. The field was his real laboratory. Although there is no evidence that he traveled in Europe, he certainly investigated the British Isles thoroughly. In addition to his studies of the Hebrides and Scotland he made many trips to England, Wales, and Ireland for the specific purpose of making observations and collecting scientific data. By his own words we learn that he had made more than thirty trips to mines. In the mines he found several minerals not previously reported from the British Isles, checked the temperature at various depths, observed the action of groundwater, including its temperature, its rate of downward migration, and studied the nature of veins, including their offset and their contact with the side walls. His many field trips, begun when he was a boy, gave him opportunities to make many geological observations.

In his work he made use of all available scientific instruments. In addition to chemical methods, he used a hand lens as well as a microscope for the study of rocks and minerals. When searching for living microscopic organisms he referred to his hand lens. On another occasion, when describing tourmaline he said that it had "minute fissures visible only to the microscope." There is no evidence that he prepared thin sections and used transmitted light, but he did use polished surfaces and reflected light. Several rock specimens with polished surfaces are still preserved in Walker's collection.

His field studies and laboratory work, begun more than thirty years before his appointment as professor, as well as his knowledge of scientific literature, prepared him for a teaching career. As will be seen from his Introduction to the lectures, he divided organized knowledge of natural history into six parts: meteorology, hydrography, geology, mineralogy, botany, and zoology. Most of what Walker included under hydrography and all of mineralogy is now included in the science of geology. It is these lectures that comprise the bulk of the present volume.

No one was required to take Walker's course; yet his classes, forty-one in 1782, fifty-eight in 1800, the last year of record, were of fairly uniform size. Even local professors enrolled. The first session of record started March 4, 1782. The second session of record started in November, 1782, and was the class in which Sir James Hall and the Rev. Mr. John Playfair were registered. The 1792 class of sixty students included, besides Robert Jameson, a Robert Brown of Aberdeen. This Robert Brown was probably the famous botanist of the early nineteenth century. (I have made no special effort to determine the possible influence of Walker upon his work.)

Walker believed strongly in the scientific method. He taught that experimentation and observation were the roads to truth. There was no value in developing theories unless they were based upon adequate observation. Again and again he told his students that adequate information is not at hand to arrive at a solution. When beginning his lecture on the "History of the Globe"[23] he said: "I will not trouble you with the theory. Burnet, Woodward, Scheill, Buffon, are all too hypothetical, and each treat of the earth's theory in their own way. Therefore, I shall content myself with making a few observations . . . confining myself with circumstances and facts."

On the value and nature of experimentation he says:

The progress of natural history has also been much obstructed, by an immoderate indulgence of theory, which has overloaded the study with much useless matter, and many useless books. . . . Nature consults no philosophers. They too seldom indeed consult her. . . . Often . . . a man conceives some general fanciful idea in natural history . . . then goes in quest of fact and circumstances to support it, and probably offers this idol to the world, by the name of his Theory or his System. . . . He is soon disqualified to be an impartial judge, either of observation or experiment. . . . Thus, a castle in the air is first formed and then a search is made for a foundation. . . . The true method of philosophic enquiry is the very reverse of this. . . . The real philosopher . . . never reasons, but from what is known. . . . In general, by means of experiments . . . new powers and properties of bodies, become thereby disclosed, and their causes discovered. The discovery of truth by experiments is the great merit of the moderns.[24]

Wherever possible he tried to use experiments in the search for truth. In 1758 he carried out more than thirty experiments in an attempt to determine the qualities of water from a medicinal well. Later, as a university professor, he explained to his students how Saussure had tried to reduce granite to whin and basalt by fire and spoke of the value of such experiments. It was probably no accident that Walker's student Hall began experiments in reducing whinstones in a reverberatory furnace in 1791, and others as early as 1790.

Walker's knowledge of scientific literature was encyclopedic. The names of more than three hundred men are mentioned in his lectures alone, not including references in his other manuscripts on botany and zoology. His own library contained more than five hundred titles, and of course he had access

[23] D.C. 2-18, Rare Manuscript Library, University of Edinburgh.

[24] Walker, *Essays on Natural History* . . . , public lecture given in 1788 on the "Utility and Progress of Natural History, and Manner of Philosophising," p. 336. Does this paragraph mean that Walker is rejecting all theories of the earth, including Hutton's? Hutton presented an abstract of his theory first in 1785 to the Royal Society of Edinburgh and published a more detailed account in Volume I of the society's *Transactions* in 1788. As secretary of the Physical Section Walker must have read Hutton's paper but probably did not recognize its importance; at least he fails to refer to it.

to the fine collection in the library of the University of Edinburgh, which is especially well supplied with ancient and medieval books. In his review of works in geology he seems to have missed very few of any importance. He told his students that if they wished to become proficient in mineralogy, they should read Wallerius' *System,* Cronstedt's *Mineralogy,* Bergman's *Scia-graphia,* Kirwan's *Elements of Mineralogy,* and Linnaeus' *Systema Natura.* None of these alone would give the whole picture, he said, but collectively they gave all that was known in the field of mineralogy. In addition, he dis-cussed some thirty-four other writers in his introduction to the subject. These included Aristotle, Theophrastus, Julius Caesar, Dioscorides, Pliny the Elder, Galen, Avicenna, Albertus Magnus, Agricola, Sibarius, Solinus, Cardan, Fallopius, Fabricius, Gesner, Cesalpino, Becker, Aldrovandus, Caesius, Boe-tius de Voel, Charlton, Thorne, Woodward, Henckel, Bromele, Newman, Pott, Lehmann, Woltersdorf, Cartheuser, Hill, Da Costa, Ludwig, and Gil-bert.

His approach to teaching was similar to the modern method. He first de-fined the subject and then reviewed the literature from Aristotle to the most recent authors. For each topic under discussion he described the phenom-enon, told what causes others had assigned to it, explained why he thought some interpretations were false, and then chose the one he thought was true or assigned a new cause, giving the pros and cons, or said that adequate ob-servational information was not available on which to base a conclusion.

A striking feature of his lectures was the use of display specimens. "In this science more knowledge may be obtained by the eye than can be convey'd by the ear."[25] Again and again he says "I now show you. . . ." His instructions to students would be as appropriate today as they were in the 1780's. "It is to be recommended to mineralogists, that they be attentive in noticing the places where a fossil is to be found or in finding out whence it came; and likewise to endeavour as far as possible to investigate the æra of its formation."

The use of printed syllabi was an important feature of his course and has remained so in geological teaching to the present. He distributed to students an outline called *Institutes of Natural History,* a classification of minerals, and a glossary of geological terms.

System *of* Mineral Classification

Walker's classification of minerals and rocks was based on the Linnaean system. Duplications and overlapping groups are readily recognized, but it must be remembered that the system persisted for many years. Walker's "Schediasma" was based on many years of field experience, a vast knowledge

[25] See text, p. 47.

COLINTON PARISH CHURCH AS IT APPEARED IN WALKER'S TIME

This photograph shows the church as it appeared before 1908, when it was extensively enlarged and reconstructed. It had changed little from the time of Walker's ministry until then. The coffin-shaped object in the foreground is an eighteenth-century mortsafe, a block of cast iron placed over a newly made grave to discourage "resurrectionists" from digging up the body for sale to anatomists. (Courtesy of the Rev. W. B. Johnston, B.D., minister of Colinton.)

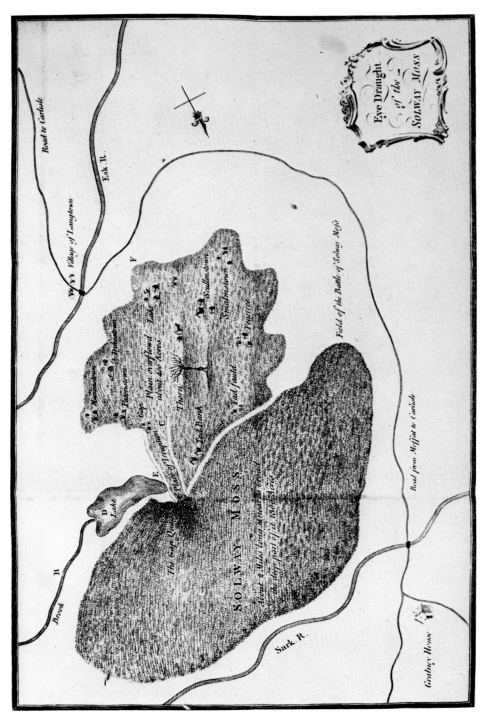

WALKER'S MAP OF THE SOLWAY MOSS SLIDE OF DECEMBER 16, 1772

This map was included with a letter to the Earl of Bute. It represents a very early map of a land-slide. Walker correctly interpreted the cause as supersaturation of the material after heavy rains.

THE BUILDING WHERE WALKER TAUGHT

The department of natural history at the University of Edinburgh occupied the upper story of the building to the right of center. This structure was completed in 1642. The building to the left, designed by Robert Adam, was under construction in Walker's time. (Reprinted from *Edinburgh in the Olden Time* [Edinburgh, 1880].)

CLASS LISTS FOR WALKER'S LECTURES

Walker's registration lists, with the record of students' payment of fees, are preserved in the University of Edinburgh library. The 1782 list above shows Sir James Hall as the first registrant and John Playfair as the nineteenth (misnumbered 20). The list below shows Robert Jameson as the eighth to register for the spring session of 1793.

of the literature, and specimens from many parts of the world. He ranks, without question, as the "Father of Scottish Mineralogy."

In building up his classification he progressed through four stages. The first was marked by an "Elementa Mineralogiae" of the 1750's and 1760's; the second by his "Schediasma Fossilium" of 1781; the third by the "Classis Fossilium" of 1787; and the fourth by a list of about 1792, with descriptions in Latin. The first was apparently developed in Glencorse and Moffat as a working base for his mineralogical studies, but no copies are extant. The second was a booklet in English for students, probably used in 1779 and 1780 and published in 1781; found in manuscript D.C. 2-19, it is included in this volume. The 1787 list included 19 classes, 67 orders, and 323 genera. "This number of genera, I believe, cannot well be much lessened, but must necessarily be enlarged by future discoveries. . . . The catalogue of my collection contains 1569 species and varieties of minerals. . . . The number of specimens . . . amount to above 3138."

In the manuscript labeled D.C. 10-33 will be found the fourth and final description of minerals by Walker, apparently an outgrowth of the "Classis Fossilium." This includes a few descriptions in English but is notable for its inclusion of about 145 in Latin. The Latin descriptions consist of three basic parts, a synonymy, "characters natural," and "characters chymical"; sometimes a fourth paragraph on "observations" is included. From the context it appears that Walker prepared the section in Latin about 1791 or 1792.

It was Walker's habit to keep his notes up to date. In the references to minerals there is an occasional reference date as late as 1790. This means only that such references were added to the original text. He says that "Since the year 1787, it has been requisite to add to this number 10 new genera."

THE MINERALOGICAL SCALE *of* HARDNESS

One of the most surprising discoveries in Walker's notes on minerals is his use of numbers as applied to hardness. Mohs published his number scale for the hardness of minerals in 1820. In the same year Jameson, after conferences with Mohs, reprinted the scale with Mohs's permission, and it has been used as a standard in mineral identification ever since.

As early as the 1780's, however, Walker was using a system of comparative hardness that was similar, though not identical, to Mohs's. Walker did not use it extensively, and the concept may not be original with him,[26] but the germ of the idea certainly goes at least as far back as Walker.

[26] Lloyd W. Staples, in "Friedrich Mohs and the Scale of Hardness" (*Jour. Geol. Education* 12 [1964]:98–101), has pointed out that Richard Kirwan in 1794 used a numbering system. Considerable research would be required to trace the history of this subject.

When discussing the gems Walker said, "Emerald . . . is harder than amethyst and softer than topaz . . . Amethyst . . . possesses the sixth degree of hardness . . . Chrysolithus . . . possesses the fifth degree of hardness . . . Topaz has the fourth degree of hardness . . . Ruby . . . is the hardest of all except diamond . . . Garnet . . . cannot be scraped with a knife . . . equal to or harder than amethyst."

It is obvious that Walker had in mind a number scale of hardness with diamond at the top. His "6" for the hardness of amethyst, and "5" for chrysolite are close to the Mohs scale, but his "4" for topaz is almost certainly wrongly transcribed, because he knew that topaz was harder than amethyst. The numbers 4, 5, and 6 are the only ones found in his manuscripts, but it seems logical to assume that he used numbers 1 to 3. From his assigning amethyst a hardness of "6" and adding that emerald is harder than amethyst and ruby is harder than all except diamond, we readily see that he had a completed scale in mind.

Walker's scale of hardness, as found in his mineral descriptions, may be outlined as follows:

Mohs scale	Mineral	Walker's Sequence and Comments
10	Diamond	Hardest of all minerals
9	Ruby	Next to diamond
9	Sapphire	Softer than ruby
8	Topaz	Softer than ruby, harder than emerald
7½–8	Emerald	Softer than topaz, harder than amethyst
7½	Hyacinth (zircon)	Harder than quartz
7	Amethyst	6th degree of hardness
7–7½	Tourmaline	Cuts glass
6½–7½	Garnet	Equal to or harder than amethyst
6½–7	Chrysolite (olivine)	5th degree of hardness
	Chalk	Hard enough to mark leather

The significance of Walker's notations on hardness is not that he was right or wrong as compared with Mohs, but that he had a concept of a scale of hardness based on numbers forty years before Mohs.

The Extent *of* Walker's Geological Knowledge

Walker's command of geological concepts can be assessed in several ways. The study of his geological and mineralogical vocabulary that appears in the

Appendix represents an attempt to compare his working concepts with those of a modern geologist. Another such attempt is represented by the following summaries of Walker's statements about a series of key concepts.

Origin of the carbonates. Walker wrote in some detail about this subject. He said:

> It is a dispute among philosophers concerning the origin of this Earth. Linnæus and his adherents say that it is purely animal, drawing its consistence from the matter of decayed shells. Others think it an original and native rock. These last allege, and this I know to be the fact, that in all our Primitive mountains and strata, our whin stone and whin rocks, granites and mountain slates, we always discover a minute portion of calcareous earth, which goes near to decide the question in my opinion, and of which opinion I confess myself. . . . Do the limestones of primitive strata yield the same products? And thus to confirm the supposition that it is formed of the exuviæ of the Ocean, this is particularly found in the strata found intermingled with shells and Corals, but the fine large alpine vertical strata, if they contain no such marine productions, and have probably neither muriatic acid nor volatile Alkali in them. In all limestones, so far as tried, there is always dissolved a portion of muriatic acid and volatile Alkali.[27]

Limestone and marble. The differences between these rocks were perhaps first recognized by Walker. On this subject he wrote:

> In all former arrangements and systems these two genera have been jumbled together, but I think that a line may be drawn which may serve to distinguish them in a generic way sufficient for the purposes of the naturalist. It is this: that all limestones are terrigenous stones or are indurated earths no ways crystallized, and hence have no transparency. Marbles are fluctivagous, their substance has been crystallized; hence, they have a degree of transparency, capable of a finer polish, and have a splendid appearance. On examining a number of limestones and a number of marbles, it will be found that there is a foundation for this distinction between the limestones and the marbles sufficient for our purpose and to preserve them as distinct genera from one another. . . . A remarkable species is the Sutton limestone found in Lancashire.[28]

Oolites. Walker was not the first to identify these, but he recognized their wide distribution in England and classified them into two major divisions based on size: pisolites and oolites. In his description of rocks he referred to "Orobias or Pisilothos—peastone . . . composed of round particles like peas . . . appears like a concretion of white peas . . . found in concentric circles. . . . Meconites—Spawn stone . . . [oolite] formed of granules . . . very minute, the particles spherical. England abounds with this stone."[29]

Caves have been studied and used since the dawn of man, but the distinc-

[27] D.C. 2-19, pp. 17 and 55.   [28] *Ibid.*, p. 63.   [29] *Ibid.*, p. 59.

tion between the kinds of cave deposits has seldom been expressed more clearly than when Walker described them. In addition to cave deposits he spoke of the work of petrifying waters and the formation of concretions.

Stalactites—water icicle, the drop-stone or stoney icicles . . . always laminated in a point; its transverse section striated is of a conical figure and is often perforated; hangs down from the roofs of caverns. This is a specimen of a stalactitical concretion. . . . Stalagma—drop stone. This genus has but a slight distinction from the stalactites; the stalagmata being formed on the floors of the caverns in which the stalactites grow. It is formed from the droppings of water from the roof; it rising upwards from the floor, while the stalactite descends downwards from the roof. . . . Sinter comprehends all those concretions which are formed from petrifying waters which contain calcareous earths dissolved in them by the aerial acid. When these waters run along and deposit their earth, undulagots are formed upon grass, wood, stone, and whatever else it meets with. Of a similar nature are the incrustations of tea kettles. Here is a specimen of it. It is a mass of shells bound together by the petrifying water of the cave of Lismore. The people of the island had a custom of resorting to this cave in bad weather and of eating shell fish and of throwing the shells in the cave. These have grown together; the petrified matter is a white chalky colour, but the water proceeds from a stratum of limestone, which is black, from whence the calcareous matter must be in a completely dissolved state.[30]

Definition of the common rocks before the end of the eighteenth century was loose and indefinite. Refinement of classification still goes on. Walker's definitions indicate that in classifying them he considered their origin as well as their physical properties. Referring to sand, grit, and gravel he said:

[sand] . . . particles water driven, polished, primarily silica, angular or round, dusty, often blown about by wind. . . . grit . . . particles are grosser and of several different substances; from siliceous earths and comprehends all the terrigenous matter. . . . gravel is an accumulation of smaller stones, tossed by waves over a long period. Nodular, worn, smooth, round, unequal in size. Varies in substance, size, shape, and color.[31]

Petroleum. The origin of oil has been a subject long debated. Walker discussed the subject and asked a very important question: "Why may there not be fossil oil?"—oil in rocks in a natural state. He posed this query almost three-quarters of a century before oil was recovered from subsurface sources. He stated that he believed that such oil occurred because "no mineral will form a soap detergent with alkalies." Apparently Walker was thinking that if any mineral would form a soap in alkali, any oil or fatty substances present in rocks in a natural state would have been washed away with the detergent action. He included oil in the inflammabilia and said:

[30] *Ibid.*, pp. 71–73.　　　　　　　　　[31] D.C. 10-33, pp. 363 and 364.

Many suppose, that they are not natural to the mineral kingdom, and think they are derived from the organized bodies of animal and vegetable substances. Sage, Cronstedt, Fourcroy, Bergman and many others are of this opinion. But as there are mineral, water, earth and acids, "why may there not be fossil oil?" It seems probable that there is, and that is inferred from the fact that no mineral will form a soap with alkalies. There are many reasons to favour the assertion, but the matter is not yet directly probed.[32]

Chemically, this is a reasonable statement. Although geologists are in agreement that oil is of organic origin, we have not yet answered all the questions on the subject. That Walker should be thinking about the origin of oil and discussing it in the classroom in the 1780's, a hundred years or so before much attention was given to the matter, seems remarkable, especially in the light of all the work that has been done in recent years.

The origin of igneous and sedimentary rocks formed one of the greatest controversies of Walker's day. It pitted the plutonists against the neptunists— by fire or by water—with Hutton and Werner the chief adversaries. Walker did not take an active part in the debate, but because the subject is of interest to geologists his opinions and observations on the matter must be called to attention.

Walker was quite clear about the origin of basalt, and could find no difference between the composition of basalt and whin rock. He never thoroughly understood the origin of sills and dikes, and early in his career he failed to recognize the evidence of former volcanic activity in Scotland. Granites, syenites, and porphyries were discussed by Walker from 1764 onward. Although he did not recognize their intrusive character, he knew their composition and said that they had been formed by fire. He had almost a full comprehension of all the different types of volcanic rocks.

Walker divided basalts into three categories: columnal whin rocks with concavo-convex surfaces and jointed, such as the basalts forming the Giants Causeway; columnal whin rock with plain joints; and columnal whin rock without joints. When discussing basalts he commented:

The stones of the Giants Causeway is [sic] composed of a fine species of this genus. It is also common in the Western Islands, forming the sides of their mountains and rocky shores . . . we have an imperfect representation of the columnar basaltes in the neighborhood in the hanging rocks in the King's Park on the Duddingston road, which are prismatical without articulations at all, but at Dunbar they are still better . . . we call it the fine black whin rock. The columns are sometimes hexagonal which indeed is the most common appearance. Some of them are pentagonal . . . there are three species belonging to this genus: as those forming

[32] D.C. 2-19, p. 317.

the Giants Causeway which are all jointed, and the juncture is concave-convex, the others are solid from end to end . . . that of the Giants Causeway and of the Isle of Staffa, and of Arthur's Seat are just of the composition and resemble the Egyptian basaltes of old. In the volcanic parts of Italy and Sicily many basaltic columns are found and seem to be volcanic . . . the common whin rocks . . . are entirely devoid of all extraneous fossils; they are often the walls of some metallic veins both in Scotland, Sweden and Germany . . . it differs from the basaltes only in being disposed in indeterminate masses.[33]

Volcanoes. Walker's knowledge of the distribution and history of Recent and near-Recent volcanoes was fairly extensive. He knew that some were extinct and that some were submarine, the latter often rising out of the ocean to form new land. He recognized volcanic glass, scoria, ash, pumice, basalt, and lava gems as products of volcanoes.

Mining. The mining of minerals is an ancient art. Walker did not contribute much to the subject, but he made many observations on his numerous trips to mines. He observed that metals occur oftenest in areas of Primary rocks and that metals are not normally found in sedimentary rocks such as coal, gypsum, or sandstone, and only rarely in limestone. He observed that a poor growth of plants may indicate the presence of a mineralized area. Much research has been carried on in the twentieth century concerning the relationship between trace elements in plants and their relation to areas of mineralization.

During Walker's visits to the Leadhill mines he observed that some veins were cut off, others offset, and that in general the number and size of veins decreased with depth. He noted that some veins were connected by intersectors or cross-veins and that small cavities were often partially filled with crystals. He believed that the mineral matter had been carried in solution by subsurface waters and precipitated in veins, often impregnating the side walls.

Types of rock structures. Recognition of various kinds of rock structures was slow in evolving. Some of the oldest terms, such as dip and strike, were used first in the mining industry. Walker included several structural terms in his lectures and used some as early as 1770 in his description of the geology of Scotland. In his lecture on the structure of the earth he stated that he believed most mountains were high on the west and dipped gently eastward. He cited the mountains of Great Britain, Norway-Sweden, the Andes, and the cordillera of North America as examples, and observed that continents and islands generally trend north-south with the long axis of the continents parallel to the long axis of mountain ranges.

[33] *Ibid.*, pp. 261–63.

Hutton made a major contribution to science when he recognized the significance of unconformities, and Playfair vividly described a trip to the field with Hutton where they observed the angular relation between two sets of strata. Although Walker did not discuss the subject in detail, he did observe in his journal of a trip from Edinburgh to Elliock about 1770 that in the Pentland Hills, Secondary strata overlapped the Primitive at the foot of the mountains. He said: "the valley at the foot of the hills abounds in sandstone, limestone, coal, blaes, dogger, and other Secondary strata. These approach the mountains, and at their feet overlap the Primitive strata."[34] Although he may have confused Primitive rocks with intrusives, he did recognize the angular relationship of beds in several places. He gave many examples of tilted and horizontal strata as early as the 1760's but did not recognize closed folds.

He spoke of horizontal strata of marl separated from the superjacent moss and said that plains are commonly formed by horizontal strata, whereas mountains are formed of tilted strata. Also, he described basalts as resting upon strata "much inclined to the horizon." He recognized the extension of strata below the ground as when he described the possible occurrence of coal from Edinburgh to Dysart under the Firth of Forth. In addition to noting the various structural relations of strata, Walker was probably the first to observe that the configuration of the surface slope of the land is directly related to the structure or kind of underlying rock.

Walker considered fractures as cracks produced by partial subsidence. He believed sedimentary rocks were formed in a fluid and passed from a soft state to an indurated state and finally fractured. He noted that the fractures were the canals along which subterranean waters or fluids might move. The fractures could thus be filled with earthy matter or ores of metals. Therefore, veins were formed after induration of the rock and were most numerous at shallow depths. Walker believed the mineral matter in veins had been formerly dissolved or diffused in water, carried from the adjacent rocks and deposited.

Glaciation. Early in Walker's career he puzzled over the origin of "bowlder fields." By 1770 he had described "till bands," and later referred to "till rocks," but could not develop a satisfactory explanation of their origin. During his excursions in southern Scotland he noted the presence of yellow chert pebbles in drift, and it was several years before he discovered their source. By the time he began lectures at the University of Edinburgh he was well acquainted

---

[34] Walker, *Essays* . . . , "Mineralogical Journal from Edinburgh to Elliock," p. 387. Walker observed the unconformable relation of the strata about 1772 and referred to it as "overlap." He did not use the word unconformity. In 1788 Playfair and Hall accompanied Hutton to the field where he showed them an unconformity and recognized its significance.

with the distribution and nature of glacial deposits in Scotland. His excellent description of "bowlder fields" includes a statement indicating that he considered the origin of these boulders and the material containing them one of the most important unsolved questions in geology.[35]

The origin of valleys was not well understood in the eighteenth century. Walker did not fully understand their origin; he believed that some valleys had been cut by the rivers flowing through them, but felt that other valleys must have had a different origin. His opinions were affected perhaps by the fact that most valleys in Scotland had undergone a complex history of glaciation, and the concept of continental glaciation had not yet been developed.

Stream erosion. Walker understood that the rise and fall of a river such as the Nile was controlled by rainfall in the area of the headwaters. He knew that tributary streams fed larger streams and that the load carried was derived from the land through which the streams traversed. The load was carried to the mouth (embouchure) of the river and deposited to form a delta. So came into existence great deltas such as those of the Nile and the Rhine. Thus new lands were formed in the sea at the mouths of rivers and deltaic soils enriched. He described watersheds as the divides between streams and stated that if they were lowered the direction of drainage might be reversed.

In speaking of the load of a stream, Walker formulated a law: "The quantity of sediment in the waters of any river varies according to the nature of the soil through which it flows and to the degree of flood."[36] How many times has this law been repeated, phrased in slightly different words, without the writer's knowing that Walker formulated it? The formulation of this law is another reason to assure Walker a place in the history of geomorphology.

Subsurface water was more or less a mystery to most scholars up to Walker's time. Walker reviewed for his students the general opinions and then proceeded to give an accurate analysis of the origin and movement of underground water. It is difficult for us in the twentieth century to appreciate the full significance of Walker's knowledge of subsurface water, especially since a certain element of witchcraft still surrounds the subject.

A review of Walker's knowledge of subsurface water shows that he was far ahead of his time. He had broken away from the past. No man before him had ever discussed groundwater so well, and it would be a long time before anyone else would equal him. He knew that subsurface water could not rise above the intake point because of the "'fountainhead"; he knew the causes of hot springs around volcanoes and the temperature control of most

[35] See text, p. 192.   [36] See text, p. 159.

springs. Walker had a fair understanding of the origin of geysers and recognized the difference between geyserite and travertine; he recognized that the intermittent nature of many springs is associated with seasons and rainfall, and that the volume of a spring may vary between rains. Walker recognized that caves and underground passages are commonest in limestone and that underground water circulation is facilitated by rock fractures and inclination of the strata, with springs issuing in the direction of inclination; that the mineral content of water varies and that mineral veins may be enriched by underground water; that seawater does not rise above sea level to displace fresh water below the land surface.

Walker had observed that the level of water in caves fluctuates during dry and rainy seasons, showing that the level is controlled by rainfall. Streams often disappear into an underground cavernous system. When mines intersect subsurface water, local springs dry up. Water percolates downward from the hanger (hanging wall) of veins and the rate of downward migration can be determined by studying mines, such as Leadhills, after rains.

The level of lakes may fluctuate between dry and wet seasons. Walker reasoned that the level of lakes was directly related to the changing level of subsurface water so that in dry seasons some lakes disappeared into subsurface caverns and channels such as many in Italy and in the Silva River area of Russia.

Mountains. Walker classified mountains into seven different types according to their geomorphic form: conical, peaked, rounded, ridgy, table, faced, and acuminated. He recognized that mountains in the process of time are lowered by degradation due to rain, wind, and frost.

He followed the proposals of Lehmann (1756) by classifying mountains into Primitive and Secondary. The term Tertiary was used only once by Walker, but he thought more in terms of Tertiary hills than in terms of a third type of mountain—more or less the manner in which Pallas (1777) and Arduino (1760) had used the term. Sometimes he used the word Primary as a synonym for Primitive. He described the Primitive mountains as composed of Primitive rocks, primarily granite, schorl, and schistus, whereas Secondary mountains are composed of Secondary strata, consisting mostly of limestone, sandstone, chalk, clay, and slate. Secondary types of strata, he pointed out, may be found dispersed in the Primitive, but never the reverse.

Walker recognized Primitive and Secondary types based on kinds of rock, age of strata, degree of folding, and presence or absence of petrifactions. Although he did not know the cause of mountain-building, other than volcanic mountains, he believed that all mountain chains had one general cause and were produced at two different "æras." Today we are still arguing over the methods of mountain-building.

Classification of sedimentary rocks. Walker believed that sedimentary rocks subsided from a fluid, strata-superstratum, one deposited upon the other by oceans and rivers over all the earth. The most ancient strata were to be found either at a great depth or exposed at the summits of mountains. Strata are not disposed according to their specific gravity. Some strata are more compact at depth than shallower strata of the same material as a consequence of compaction. Unconsolidated sediments in valleys, on plains, on seashores, along the banks of rivers and at their mouths, he classified as Accidental. Under Accidental Walker included all soils, deltaic deposits, shoreline and terrace sediments, and surficial materials now known as glacial drift. This would encompass all of the Recent and Pleistocene unconsolidated rocks.

Peat is one of the most important soils in Scotland and the Hebrides. Walker's interest in agriculture and botany led him to give serious attention to the origin and nature of peat as early as 1762. He had described the Solway Moss flow, a landslide, in his second published paper. His lectures on the origin of peat represented an outline of what was included in "An Essay on Peat" published in 1803. His students, including Jameson, were well informed on all phases of peat development and use. Jameson's later interests in peat and kelp were natural developments from the teaching he had received.

Walker's essay on peat is perhaps one of the classic papers on the subject. Here he denied that peat is marine in origin, remarking that when the sea throws up peat in Holland it means that the sea has encroached upon the land and exposed the peat to wave action. He classified peat into seven different types but later reduced the number to four. His major conclusions were: (1) the texture of peat depends upon the degree of putrefaction, (2) all peat has been formed where it is found, (3) it is composed of half-putrid vegetable matter, formed by the decay of plants in the place where the plants grew, and (4) peat is the peculiar product of temperate and cold climates and cannot subsist in warm climates because of the degree of putrefaction. He identified eighty-four species of plants commonly found in peat and mentioned that many more probably exist.

Earthquakes. The origin of earthquakes remained somewhat of a mystery to Walker. He knew that one class was associated with volcanic eruptions, but the other class was a complete enigma. He would not accept them as connected with mountain-building. His attempt to associate their frequency with seasons shows that he had not comprehended the cause of the second class.

Temperature stratification in lakes. Various explanations have been given for the lack of freezing of some lakes in winter. Walker attributed the phenomenon to the temperature stratification of the water. In deep lakes the water at a certain depth maintains a constant temperature above the freezing point. He was probably the first to recognize a seasonal stratification of lake water.

Organic fossils. The significance of fossils was not well understood in Walker's day. Although he wrote little on the subject, what he wrote shows that he had a superior knowledge of the basis of classification. He referred to petrifactions as extraneous fossils or simply as petrifactions. He classified the methods of fossilization under six headings: induration, incrustation, concretion, coagulation, crystallization, and insertion. It is not clear whether he meant by insertion the process of replacement or the filling of void spaces. This probably represents the first attempt to delineate the methods of fossilization.

His ideas of the criteria of classification were almost as clear as ours are today.

[1] Species, in Philosophical language means such objects as are subordinate to genus.

[2] Varieties are those beings belonging to any species and differing from it in some trifling circumstances . . . nature is slow in producing these changes and requires time and circumstances to effect them.

[3] Varieties in the Animal Kingdom proceed no doubt from different climates. . . . We know that Climate and pasture when joined together produce a remarkable change in our Domestic Animals.

[4] Some varieties in animals are hereditary and permanent . . . in some animals a variety will miss one generation and will take place in the next.

[5] Hybrids or mules are produced by the conjunction of two different species of plants or animals.[37]

Walker was puzzled by the absence in Great Britain of any living representatives of species of fossils found in the chalks. He did not understand or recognize the extinction of species even though he knew that "organized bodies of animals or vegetables are found in almost every part of the globe petrified." He mentions petrified amphibia, fishes, insects, vermes, including gastropods, *Ceratites, Belemnites, Echina,* madropore corals, and *Ammonis.* He said: "By far the greater part of the shells of Britain are not now to be found in a recent state. . . . *Cornu ammonis* so plentiful in England to the number of thirty-nine species has never been in recent or living state."[38]

William Smith (1796) originated one of the most far-reaching concepts of geology when he recognized the vertical change in fossil faunas, and that the

[37] D.C. 2-20, pp. 15–18.   [38] *Ibid.,* p. 5.

relative age of rocks could be determined by fossils. Walker was close to expressing the same idea when he said:

> These extraneous fossils are useful to ascertain what changes the earth has undergone, and it is a question how these various matters have got to the tops of the greatest heights and to the depths of the profoundest caverns of the earth . . . the proper method of investigation in these subjects is not the analytic hereto pursued by conjectures but the synthetic which alone must conduct us in the present limited state of our knowledge for facts and data [are] yet wanted to build theories upon . . . first of what year to fix its formation, second, *how far its Chronology can be ascertained by the physical data, for these are very probably certain monuments by which we may Judge this matter.*[39]

This is one of the earliest suggestions in geological literature proposing that fossils may be used in determining time chronology. Again, Walker was far ahead of his time, and this proposal alone would mark him as a man of great talent.

Various theories have been proposed to account for the abundance of life in Cambrian times and its paucity in the Precambrian. Before recent discoveries of fossils in Precambrian sediments most geologists reasoned that the abundance of highly developed invertebrates in the Cambrian necessitated a long Precambrian history for the diversification of life. Recent discoveries of well-preserved fossils in Precambrian sediments confirm the existence of highly developed invertebrates before the Cambrian. Walker used the same basic argument because he believed that the existence of life in the oldest Secondary strata required the presence of life on earth in the Primitive. He believed that the various forms of life were transitional and spoke of "a continued chain in nature from its lowest subject up to the human species."[40]

AN EVALUATION

Walker was certainly the "father of Scottish mineralogy" and he was certainly the "father of geological teaching." Furthermore, I believe that Walker must be regarded as the first man who can properly be called a geologist, others before him having been for the most part chemists or mineralogists. His botanical manuscripts may assure him a place as one of the eighteenth century's leading botanists. And when his agricultural papers are fully evaluated, they will surely place him as the leading Scottish agriculturist of his day, if they do not accord him a still higher place in history as the "father of agricultural education."

[39] *Ibid.*, pp. 2 and 3 (italics mine).   [40] See text, p. 23.

Walker *the* Man

Geologist, botanist, agronomist, Presbyterian minister—one wonders what sort of man John Walker really was. His preoccupation with natural history might make it appear that he was a poor minister, but just the opposite seems to have been the case. At least one of his parishioners was inspired to write a poem eulogizing him in the most bombastic terms. "Hail, Walker, hail! this champion ever prove."[41] More significant is the fact that in 1790, the General Assembly of the Church of Scotland appointed him Moderator, the highest office the church could bestow. All the facts indicate that Walker was an outstanding churchman as well as a remarkable scientist.

After his appointment as regius professor he was assigned to the parish at Colinton in 1783. The beautiful church tucked away in a heavily wooded vale was much to Walker's mood and temperament. He especially enjoyed the cultivation of willows, his beloved *Salix.* Here he could work in the surroundings that later would inspire Robert Louis Stevenson to write of his childhood home, "Even at the Witche's Walk, you saw the manse facing towards you, with its back to the river and wooded bank, and the bright flowerpots and stretches of comfortable vegetables in front and on each side of it . . . a well-beloved house—its image fondly dwelt on by many travellers."[42]

Not much is known about Walker's social life. In addition to participating in the morning levees and dinners he must have been involved in many church functions. His social obligations may have been somewhat limited during the greater part of his life because he was a bachelor until, in 1789 at the age of fifty-eight, he married Jane Wauchope of Niddry.

In regard to Walker's work habits Tytler reports that "it was his custom, for a great part of his life, to indulge himself in nocturnal study; seldom feeling the resolution to quit his books and papers until four or five o'clock in the morning, and of course passing the best part of day in bed, a practice which destroyed a good constitution, and in the end was attended with a total loss of eye sight, for the last six or seven years of his life."[43]

Walker died in Edinburgh December 31, 1803, and was buried in Canon-

---

[41] F. J. Guion, "Poem to the Rev. Mr. Walker," *Scots Mag.* (1772), pp. 372, 441.

[42] Graham Balfour, *Life of Robert Louis Stevenson* (New York: Scribner's, 1901). The first part of the quotation comes from "Reminiscences of Colinton Manse," unpublished manuscript at Yale; the last line comes from Stevenson's "The Manse," second paragraph. Personal communication from Bradford A. Booth.

[43] A. F. Tytler (Woodhouselee), *Memoirs of the Life and Writing of Honourable Henry Home of Kames* (Edinburgh, 1807), p. 12.

gate Cemetery. His wife survived until May 4, 1827. On the headmarker we read:

The Burial Place of the Revd. Dr. John Walker,
Minister of the Gospel
Colington
and Proffessor of Natural History
in Edinburgh

In a memorial on Sir Andrew Balfour, Walker wrote his own epitaph better than anyone else. In describing Balfour he used words that I suggest are equally fitting for himself:

Nor was his character less conspicuous as a man, than as a scholar. Pious and virtuous; noted for the candour and generosity of his mind; for his benignity, munificence and public spirit; his whole life seems to have been a series of public and private offices to God, to his King, to his country, and his friends; and he seems, in fine, to have possessed all the virtues that enter into a great and good character.[44]

[44] *Essays on Natural History,* p. 364.

CONTEMPORARY CARICATURE OF WALKER BY JOHN KAY

INTRODUCTION

I cannot proceed without a good deal of diffidence when I reflect upon the difficulties that occur and the disadvantages I labour under. I am to teach a science I never was taught. In the other sciences we are here regularly instructed, the Lectures of Professors we can obtain and follow. But in Natural History we have not here this advantage, for it never has been taught in this University.

In Lecturing therefore on a new subject, and on one consisting of such a multiplicity of different parts, I cannot pretend to appear every where entertaining to different persons. I am likewise sensible I shall often be liable to appear trivial to some and perhaps obscure to others.

Tho the History of Nature has been the subject of my amusement from my younger years, and tho it appears perfectly familiar to me, yet when I come to take a more comprehensive view of the whole, with a design to teach it to others, things appear in a different light. In this situation I believe every person will find more reason to be surprized at the imperfections of the science than at all the labours the learned have bestowed upon it.

A complete system of natural history does not exist any where, it is not to be collected from books, there are many blanks to be filled up, as many objects of it are still buried in obscurity and there are many new parts of such a system which have never been enquired into, or but little cultivated. For these difficulties naturally follow from its having been prosecuted in detached parts which have never been altogether collected into a regular science.

These several disadvantages make me see how much I stand in need of the candour and indulgence of those who favour me with their presence; and these have all along prevented my being certain as to the success of my present undertaking.

During the survey of natural history I have had in reflection I am surprized that so many branches of useful and ornamental knowledge should in this place never have been the subject of public instruction. Among men of education scarcely a day passes in which some part of natural history does not occur in conversation, and yet few know any thing of it but what they have reap'd from books, or the conversation of others.

I look upon Natural History as necessary to form an accomplished Gentleman tho' too much neglected in their education, and I likewise esteem it necessary to form an accomplished Physician. And I believe it will be found that from the first beginning of physic, in the middle and in the modern ages the most eminent have always been remarkable for their skill in natural history. In other Classes the audience have usually some general knowledge of the subject, they know in some degree the general nature and extent of the science, but here for want of opportunity we have no such previous knowledge.

The term natural history is commonly used so indefinitely that I believe, it must be difficult to know what must be taught in any treatise on the subject. It is therefore highly requisite to give a Definition of the science, to mark its limits and to point out in general the subject of our present course. But pursuant to my general custom I propose making mention of the Discoveries and improvements that the last year has afforded in our science.

DISCOVERIES *and* IMPROVEMENTS *in* NATURAL HISTORY *Made in* 1788

The translations from the works of Linnæus have been very numerous, but their merits do not entitle them to a particular enumeration as they are all inferior to the original.

Many Philosophers on the continent have employed themselves with great ingenuity in endeavouring to construct Hygorometers [sic][1] on established principles, but it is to be regretted that in the formation of these we cannot ascertain any fixed points so as to give the Instrument the stability and precision of a Thermometer.

In the Thermometer, when the great Sir Isaac Newton first considered

---

[1] Hygrometer—an instrument for measuring the humidity in the air.

the subject he discovered three fixed points, vizt. the freezing point of water, the heat of the human blood, and the point at which water boils, to these three, succeeding observations have not been able to add a fourth. The boiling point has been long known to be regulated in some measure by the pressure of the Atmosphere, but Monsr. Rushard[2] of Berlin asserts from the result of many ingenious experiments that it is also influenced by other circumstances, but which he does not pretend to explain. I however agree with him in supposing that it may be found by future observation to depend on some, as yet undiscovered, law of nature.

To determine the salubrity of different parts of the Atmosphere by ascertaining the quantity of pure air, it contains, the Abbè Fontana[3] and Dr. Priestley[4] recommend its mixture with nitrous air; Mr. Scheele[5] with Inflammable Air; and Mr. Ruchard doubting the precision of either advises the deflagration of Kunkel's phosphorus, but his method is certainly liable to the same objections as either of the former.

Dr. Rammell of Provence has published a book[6] in which he endeavours to prove that the diseases which have been usually supposed to depend on some hurtful property of the Atmosphere are only to be attributed to incidental circumstances such as Regimen; Temperature, passions of the mind etc.

Another publication of the Continent asserts the propriety of an old idea vizt. the periodic return or Cycle of seasons, but contrary to the ancient opinion, it is limited by this Author to 18 years. However I can from my own experience assert that the years 1757 and 1740; 1741 end 58; 62 and 79, 66 and 83, which according to the position of this Author should be similar, were perfectly opposite in Temperature.

The Formation of Ice has been generally supposed to begin at the surface and afterwards to extend downwards. But Mr. Pott[7] has found that many of the Rivers in Germany begin to freeze at the bottom, and form what the Germans call ground Ice; it has often been discovered on drawing up Fishermens nets, and even Anchors when no congelation took place at the surface. Mr. Pott has also observed the same appearance by droping nails

---

[2] Or Ruchard (unidentified).

[3] Felice Fontana (1730–1805), Italian naturalist and professor of mathematics in Florence, developed a type of eudiometer.

[4] Joseph Priestley (1733–1804), chemist.

[5] Carl W. Scheele (1742–86), Swedish chemist.

[6] M.-F. B. Ramel, *Consultations de médecine, et mémoire sur l'air de Gemenos.*

[7] Johann Pott, *Lithogéognosie.*

and other extraneous substances into water placed in a freezing mixture; he prevented the surface from congealing by agitation, but Ice was formed at the bottom on the different substances he threw in.

Water has always been supposed to assume a Hexahedral shape in its congelation, but Mr. [Romé] de Lisle[8] has found that it puts on an octahedral form which was also observed in the great Hailstorm at Paris, but these instances are not sufficient to overturn the generally received opinion as we know that the Crystalization of every salt (and as such we must consider water, when freezing) is much varied by the slightest accidents.

Another late publication teaches us the manner of preventing the Choaking up of Harbours by Sand carried down by rivers etc. The Authors mode of preventing this is to increase the velocity of the current by narrowing its banks. He dignifies his work by the appellation of a Theory, a name now frequently perfix'd to the most trivial observations.

The contending Chemical Theories seem to claim almost equal belief. The Phlogistians[9] under the banners of Mr. Kirwan[10] assert after Stahl[11] the existence of a principle of Inflammability which Mr. Kirwan supposes to be inflammable Air. Dr. Demens[12] thinks that Phlogiston is a combination of Inflammable air with a salt either Acid or Alkalie. The Antiphlogistians at the head of whom is Monsr. Lavoisier have discarded the idea of an Inflammable principle and have explained the appearances which were supposed to indicate its existence upon different principles. The principal fact that supported this doctrine was the formation of water by the combination of two Airs, which was supposed to be synthetically proved by the experiments of Monsr. Lavoisier and Mr. Cavendish; but has been since controverted by Dr. Priestley whose conversion to phlogistic principles forms a striking event of the last year.

He supposes that the water produced, existed before combustion in the Airs, and that the real product of the two Airs is nitrous acid. I must own that I find it difficult to determine which theory is best entitled to our belief, but I am upon the whole inclined to agree with Dr. Priestley in supposing water to be an elementary principle.

Fluor acid in consequence of its property to dissolve glass is employed for Etching on that substance, as the nitrous acid is on copper.

[8] Jean Baptiste Romé de Lisle, *Essai de crystallographie.*

[9] Those who believed that fire was a substance.

[10] Richard Kirwan, chemist and mineralogist, member Royal Irish Academy.

[11] G. E. Stahl (1660–1734), a German, perfected the phlogiston theory.

[12] Or Demons, possibly J. B. van Mons.

Phosphoric acid was first discovered in an ore by Dr. Gaun;[13] it has been since found in cubic Chrystals in Switzerland and Mr. Prusti[14] has found it in large quantities combined with calcareous earth.

From Boracic Spar whose properties are well known the Sedative Salt has been obtained in its pure state and it is supposed that it will be found by future observation to be much more copiously dispersed than it is now believed to be.

The discovery of the Sedative Salt and Phosphoric Acid or crystallized fossils[15] gives us reason to hope that all the Gems and other Chrystallizations may be perfectly analyzed, and also verifies the maxim that I have been long inclined to hold. *Omnis Crystallizatio a Sale.*

The Botanical improvements are considerable. The *Flora Danica* still continues to be published under the inspection of a man fully acquainted with Botany.

Monsr. Pirreaux[16] is engaged in a desirable pursuit, a description of the Pyrrenean plants.

Determining the Species and varieties of plants has been attempted by Monsr. Necker,[17] but from his former fancifull publications we cannot suppose him fully qualified for such an arduous undertaking particularly as the Celebrated Linnæus has before acknowledged himself unequal to the task.

In Zoology we have also some new discoveries. Sparman's[18] Museum still continues to be published in Fasciculi at Stockholm.

In the *Systema natura* of Linnæus the Species of Animals of the Class mammalia were confined to 218 but in a subsequent publication they were reckoned to be 322, and are now in a German publication computed at upwards of 400.

The African animal the angora goat that furnishes the article of manufacture called Mohair has been imported into France and Spain, where after continuing for a long time it has not been found to degenerate in the smallest degree.

[13] Johann Gottlieb Gahn, chemist, professor at Bergcollegiums of Stockholm.

[14] Perhaps Joseph L. Proust (1755–1826), French chemist.

[15] Fossil was used in a broad sense to refer to anything dug from the earth, including rocks and minerals. Organic fossils were referred to as "extraneous fossils." The reader should usually substitute "rock" or "mineral" when reading the word "fossil."

[16] Unidentified.

[17] Noël Joseph de Necker (1729–93).

[18] Anders Sparrman (1748–1820), professor at Stockholm.

IMPROVEMENTS *and* DISCOVERIES *in* NATURAL HISTORY *in* 1790

The advancement of liberty naturally keeps pace with the advancement of Literature; in all ages and in every country they have risen and fallen together.

The advancement of literature in this present age has been promoted by natural history; it was the improvement of Natural knowledge that laid the foundation for the Royal Society of London which is the parent of all the other literary Societies in Europe. Of all the branches of science there is none that connects and combines mankind more than natural history.

Of late years we have had a Royal Society founded here. A Royal Society was also founded in Ireland in the year 1787 or 88.

A Philosophical Society has been lately founded in America. I am informed that in America natural history is making rapid progress. Dr. Barton who was not long ago a student here is appointed Professor of Natural History at Philadelphia.

There has lately been founded an Academy of Sciences at Lisbon a place where we could le[a]st have expected it.

Another at Crilut[?] in Dalmatia on the confines of Turkey. A new society has likewise been instituted at Dronthon;[19] another in Iceland and a very prosperous one has been instituted at Bengal, another at Cape Francois, and what is still more surprizing one in Mexico.

From all these we have reason to expect great advantages to Natural History.

I shall observe the same order in giving an account of the improvements, that I follow in this course.

First as to

METEOROLOGY

the subject of the Hygrometer has come of late to be throughly canvassed.

We have now four different Hygrometers, planned by 4 different Professors, from all which it is to be hoped that an accurate instrument of this kind will be soon discovered. The defect of the Hygrometer is a want of fix'd points, but I have reason to believe that this may be brought to accuracy.

There is a physician at Aix in Provence, a Dr. Reynolds[20] who controverts the opinion of the most celebrated Physicians, upon the connection between

[19] Trondheim?   [20] Unidentified.

some diseases and the qualities of the Atmosphere, and is of opinion that they do not at all depend upon the state of the Air, in this opinion however it is not very likely he will be followed.

Dr. Keir[21] in the Chemical Dictionary, scarce yet published, has found that in Atmospheric Air the pure and phlogisticated airs are not merely mix'd but are Chemically combined. The reason of this I suppose is from two facts, Ist. they are always known to exist in the same proportion in Atmospheric Air; IIdly. they are different in point of Specific Gravity, consequently, if merely mix'd they ought to separate from one another.

Of late some curious observations have been made on the communication of Heat by Colonel Sir Benjamin Thompson; this Gentleman has made the comparison betwixt the conducting power of common Atmospheric Air, and the Torricellian Vacuum, and has fix'd this position "that common Atmospheric Air conducts heat in proportion to the Mercurical vacuum as 1000 to 605"; he has discovered another principle "that the conducting power of Air is greatly increased by Humidity"; (this fact may probably be of use in medicine) hence the great coldness of moist air; and hence also the pernicious consequences of damp houses, damp beds and falls of dew. He has likewise found the conducting power of Mercury to be to that of water as 1000 to 313. Hence the reason of the greater sensation of Coldness by applying the finger to Mercury tho a Thermometer shows it to be the same temperature as the surrounding Atmosphere.

During last winter Quicksilver has been frozen in England by means of melting snow with the concentrated fuming spirit of Nitre, it has thus been made malleable and it sinks in fluid Mercury.

In the Berlin memoirs several Experiments are related which were made to determine the relation between the quantity of salts dissolved in water and the increased bulk of the fluids, from which the opinion is clearly related "that the particles of the salt occupy spaces between the particles of the water.["]

With regard to

Mineralogy

This seems at present to be the most cultivated part of Natural History in every part of Europe, and indeed it is the part most behind. There have been many considerable and ingenious disputes about the primitive Earths. The number of these has been varied by different naturalists. Bergman[22] settles them at five and I think it will not be easy to enlarge or diminish that number.

---

[21] James Keir, *Dictionary of Chemistry.*   [22] Torbern Bergman, *Sciagraphie.*

Mr. Crell[23] demonstrates that all the five primary earths are found in the primary Mountains and that no others are to be found in the secondary ones.

Mr. Klaproth[24] analyses what appears to be a new species of Gem called Zyrcomes; a hundred grains of this contain 31 of Siliceous Earth, one half grain of Iron and as much of Nickel, but the most remarkable circumstance is that the remaining 68 consist of what appears to be an Earth *Sui generis,* soluble in Acids but not in Alkalies. This gem is the Jargon of Ceylon,[25] a species of the Diamond, it is not unlikely that this substance is nothing more than a modification of the Gemmeous earth.

Mr. Georgie[26] of the Prussian academy has combated the supposed vice versa transmutation of Calcareous and Siliceous earths into one another.

—— of Vienna some time ago, published a work in which he supposes that the five primary earths may be transmutted into one another. Bergman could never do this by artificial means.

It has been supposed that Calcareous Earth may be changed into siliceous because many of the Calcareous Earths have been found to strike fire with Steel. Mr. Georgie examined 14 Prussian marbles. He found Siliceous earth in them all but one; they all struck fire with Steel, he found the siliceous earth but in small quantity, and they were all combined with Iron which rendered them hard and hardness is the only thing required for Scintillating.

Rocks of the primary kind contain a small proportion both of Argillaceous earth and of Magnesia. Mr. Georgie has confirmed a remark that has been made, i.e. that in Marbles of the primary kind there has not been found the smallest vestige either of Muriatic Acid or Volatile alkali; on the contrary all marbles and Lime that belong to the second Class show evident symptoms of both.

Mr. Beaumè[27] a good many years ago maintained that Siliceous and Argillaceous earths were convertible into one another; this has been strenuously and successfully refuted by Wolfe[28] and Bergman.

One remarkable instance alledged in favour of this doctrine is that the

---

[23] D. Loren von Crell, *Chemisches Annalen* (1784); translated three volumes of Kirwan's *Mineralogy* into German in 1796.

[24] Martin H. Klaproth, *Mineralogical and Chemical History of the Fossils of Cornwall.*

[25] Colorless variety of zircon.

[26] J. G. Georgi, *Bemerkungen auf einer Reise im Russischen Reiche im Jahre 1772* (St. Petersburg, 1775).

[27] Antoine Beaumé (Baumé) (1728–1804), French chemist, inventor of Baumé hydrometer. Works include *Chimie expérimentale et raisonnée* (3 vols.; Paris, 1773).

[28] C. F. Wolff (1733–94).

earths which are thrown out by Volcanoes being exposed to Mephetic exhalation fall into what appears to be a mere Argillaceous Earth. In the Royal Society of Sienna; a premium has been offered to ascertain whether a real change of Siliceous into Argillaceous earths takes place or whether the two Earths still retain the same proportion.

There is a remarkable species of marble in a palace at Rome, of which the first notice we had was from Mr. Ferber,[29] called the Petra Elastica. Six slabs of it are preserved one foot in length which if laid across a bar over a Table becomes convex and the two ends touch the Table.

Last year Baron D——— produced to the Academy of Sciences of Paris a remarkable Stone from the Brazils which was elastic, but what was remarkable it was of a Quartzy and siliceous nature. It struck fire with Steel, and contained a considerable portion of mica from which it acquired its flexibility.

Mr. Hassenfratz[30] has found a peculiar white earth, the natural Phosphorate of Lime, in Hungary. This earth is phosphorescent[31] when thrown upon red hot Coals. It has been lately discovered in Spain, some entire hills are composed of it. This Earth it is supposed is derived from the Bones of Animals. It is got from the hills of Estramadura, it leaves its Phosphorescent appearance after it has once or twice been thrown on red hot coals.

An extraordinary stone has been lately discovered in the Dutchy of Burgundy formerly known by the name of Calcareous Quartz. It has partly been examined by Mr. Meyer,[32] who has found that it is composed of small Crystals, not Cubic but Polygons, having 26 faces, sometimes white; grey and of an amphistian colour, subpellucid, striking fire with steel and cutting Glass. The composition truely marvelous a hundred parts contain 68 parts of the acid Borax, 13 of the earth of Magnesia, eleven of calcareous Earth, two of Siliceous and one of Argillaceous with a small portion of Iron.[33]

It is called calcareous salt of Borax and is found in cliffs of Gypsum rocks that are stratified.[34]

Mr. Saussure[35] received a present of a fossil from the Duke of Gordon,

---

[29] Johann H. Ferber, German mineralogist.

[30] Jean Henri Hassenfratz, *Méthode de nomenclature chimique.*

[31] Thermoluminescence is the phenomenon indicated.

[32] J. C. F. Meyer, chemist of Stettin.

[33] This mineral seems to be one of the complex borates, possibly related to hydroboracite or indeborite.

[34] The significance of "layered" or "stratified" rock is clearly understood in this case as in many others.

[35] H. B. de Saussure, *Voyages dans les Alpes.*

found in Banffshire, examined by Mr. Saussure Junior. It was accompanied by a note that it was got from an old miner who laboured much in this country in the reign of James the 6th.

By the analysis given of it, it must be referred to the tribe of Schorles,[36] by no means the Sappar of Cornelius. The stone he called Sapparius was found in Crawfordmuir, and is blue Crystals of Copper. It was the discovery of this that gave rise to the rich lead mines that have existed in that country ever since. The name sappar seems to be a corruption of Sapphire.

Mr. Wiegland[37] has analized a species of green g[ar]net, 100 grains contain 26 of Siliceous earth, 30 of Calcareous, and 28 Gemmeous.

Messieurs Bergman, Weigland and Kirwan have varied upon the composition of the common feldt spar.

There is a remarkable stone that has been examined by Mr. Clupper,[38] of which I had some specimens that were brought from India. It is known by the name of Adamantine spar,[39] and is a very singular kind of stone, there are two species. It is next in hardness to the Diamond, it resists all decomposition to a wonderful degree. It is composed of Argillaceous earth and another unknown earth; it is totally insoluble in acid and cannot be vitrified without alkalies.

Mr. Fourcroy[40] has properly distinguished between Siliceous earths and the gemmeous. Common Siliceous and Quartzy earths run into Glass with fix'd Alkali; but the gemmeous requires Borax or the microcosmic acid.

The same Gentleman who has analyzed the Adamantine spar has also examined a new species of Crysolite[41] brought from the Cape of Good Hope, called by the name of Preehint;[42] he also examined minutely another gem, the Crysoprack.[43] Of 300 grains he found that 208 were siliceous earth, the greatest proportion that has ever yet been detected in any other stone. In this there was only 74 grains of Argillaceous earth, whereas in several other gems, the argillaceous earth is in greater proportion than the Siliceous;

[36] The term schorl was used by Da Costa in 1761. It most commonly refers to tourmaline schist, but other rocks may have been indicated.

[37] Perhaps Johann Christian Wieglieb (1732–1800); his papers on chemistry include *Geschichte des Wachsthums und der Erfindungen in der Chemie in der neuern Zeit* (1790–91). In this line Walker's spelling of garnet is *granet,* a variant of the older *grenat.*

[38] Unidentified.

[39] A synonym of corundum.

[40] Antoine Fourcroy, *Elements of Natural History.*

[41] Chrysolite.

[42] Prehnite (announced 1783). See A. G. Werner, *Berg. Jour.,* I (1789).

[43] Probably beryl.

there were two grains of Calcareous earth, $\frac{1}{4}$ grain of Iron and 3 grains of Nickel, these fractions give the appearance of a most surprizing minuteness.

We come now to the

Saline Class

The Acid of Borax has been now discovered in a separate state. Mr. Wintler[44] found it where it was never suspected before, in the Petroleum of Hungary. It has now been discovered in a variety of Fossils and is found to be more common than has hitherto been imagined.

A French Chemist, Mr. Chaptal[45] has written a paper on what he calls the vegetation of Salts. When salts come to Crystallize they shoot briskly along the surface. Mr. Chaptal has discovered that this shooting is greatly checked by light and promoted by darkness.

Some remarkable discoveries have been made on the subject of Phosphoric acid. It was first discovered by Mr. Homberg[46] in human urine and afterwards in the bones of Animals. Mr. Margraaf[47] detected it in several vegetable substances but especially mustard, and other tetradynamious plants. It is probable that it has been always ultimately derived from the organized bodies of nature.

Dr. Gaun[48] of Stockholm first found it in the fossil kingdom in the Crystallized ore of lead.

Mr. Meyer found it subsisting in Sideritis; this turned out nothing more than a combination of Iron and Phosphorus; from this there is reason to suspect that phosphoric acid is present in the Iron Stone.

The fossil alkali is a rare production in Europe but it has lately been found in several Caverns in Switzerland combined with a large proportion of Glauber salts.

Mr. Fontana found Epsom salts in great abundance in a quarry of Gypsom.

The ingenious Mr. Saunders,[49] a surgeon at Bengal lately went upon a journey to Tiviot[50] which has always been known to be the only place that

[44] Perhaps Johann H. Winkler (1703–70), professor of physics, Leipzig.

[45] Jean Antoine Chaptal, French chemist, *Eléméns de chymie* (1790).

[46] Wilhelm Homberg (1652–1715), chemist, published several articles in *Mémoires* of the Paris Academy between 1692 and 1714.

[47] Andreas S. Marggraf (1709–82), German chemist.

[48] See n. 13.

[49] Perhaps William Saunders (1743–1817), College of Physicians (who lectured in chemistry and pharmacy).

[50] Tibet (Thibet)?

produces borax. Mr. Saunders found that the Tincal was not refined by boiling as has been commonly imagined, but brought to Market precisely in the same state in which it is dug out of the Lake.

We next proceed to

Inflammable Fossils

In the Petersburgh ads Mr. Lachelf[51] has shown that sulphur is capable of being fixed; it was Crystallized with the Aerial Acid.

Dr. Anderson[52] has visited several of the [West] India Islands. He has found in the island of Trinidad a plain of Bitumen three miles round; in hot weather it is quite fluid for one Inch in depth; he thinks it is the same kind of Bitumen that prevails near the Dead Sea in the Land of Judaea.

In 1748 the Master of the mines at Cronstadt discovered a native regulus of antimony. It is however rare. Dr. Hunter[53] had one Specimen of it in his museum that cost him 30 Guineas.

A mineral has been discovered at Facebay in Hungary composed of Antimony, Pyrites and some Gold, and always the Antimony, is in the form of Regulus.

It is not long since the saline nature of Arsenic has been established; it is soluble in 15 times its weight of boiling water. Mr. Lacholf[54] established this conclusion that Arsenic is naturally in a dissolved and fluid state.

Mr. Klaproth has discovered what he calls a semimetal in Pitchblend; it is an ore that contains lime and a small portion of Silver mineralized by Sulphur. Mr. Klaproth found that it affords a metal *sui generis* which is exceedingly difficult to reduce. It gives to Porcelain a yellow colour, its specific gravity is as 6.440. He calls it Urinite [Uranite] in a letter to Mr. Proust.[55] It is an ore mineralized by Arsenic.

With regard to

BOTANY

Mr. Koelreuter[56] at Petersburgh performed of late some accurate experiments in order to produce a plant from two different species. He performed

[51] Mr. Lachelf is unidentified; possibly "ads" refers to the satirical journal *Adskaia pochta* (St. Petersburg, 1769).

[52] Perhaps James Anderson, *Letters to Sir J. Banks* (Madras, 1787–88).

[53] John Hunter (1728–93), founder of the Hunterian Collection, Royal College of Surgeons, London.

[54] See n. 51.

[55] See n. 14.

[56] J. G. Koelreuter, known for experiments in hybridization, *Das entdeckte Geheimniss der Cryptogamie* (1777).

156 experiments upon *malva's; napra, althaea's* etc. by endeavouring to impregnate the pistillum of one species with the pollen of another. In these experiments there were no ripe or fertile seeds produced. But in 16 other experiments he seems to have succeeded upon 6 species of *Lavatera, Malva, alcea* etc. with those plants to which they seemed to be most nearly and most naturally allied. They produced seeds which produced plants partaking of the nature of the Father and Mother plants.

But here lies the misfortune. I am quite clear that the plants on which his experiments succeeded are of the same species and that they are only varieties, tho delivered by Linnæus as different species. The same thing I happened to observe in an Experiment made by the late excellent Dr. Solander,[57] they were only varieties of the same species.

From this it would appear that the production of hybrid plants requires more experiments before any conclusion can be found.

Mr. Bergius[58] has given us a new genus of *Felix* which he calls *Senoctus*.

Dr. Sibthorp,[59] Professor of Botany at Oxford, has returned from a long per[e]grination in the Levant and has brought with him about 200 species of plants that have escaped all other Botanists, and he has been at a great deal of pains to settle the synonima of the ancients.

Two Spanish Gentlemen have lately arrived from Peru and I am told have brought with them 2000 species of plants, scarce any of them known to the European botanists. They brought 71 species of Arboreous plants alive.

A curious incident occurred to Mr. Hagrin[60] in Sweden. He observed a sort of Gleam of Light on the surface of the common marigold (*Calendula officinalis*). It appeared in Sweden in the months of July and August last about sun set and for half an hour after; the air was serene and free from vapours. It appeared on the same flower twice or thrice successively like flashes. It is more remarkable upon those of a deep orange colour than those of a light yellow.

He found the same appearance also on the *Tropæolum majus* and upon

---

[57] Daniel C. Solander (1736–82), Swedish student of Linnaeus, was on Captain Cook's first voyage.

[58] Peter Jonas Bergius (1730–90), Swedish botanist listed in G. A. Pritzel's *Thesaurus literaturae botanicae*. *Senoctus* is unidentified.

[59] John Sibthorp (1758–96).

[60] Lars Christoffer Haggren (1751–1809), *Kungl. Svenska Vetenskapsakademiens hendlingar* (Proc. Roy Swedish Acad. Sci.), Vol. 9 (1788). Identified by W. Odelber, librarian of the Royal Swedish Academy of Science, Stockholm.

the *Lilium bulbiferum* (orange Lily). He likewise observed it on the Sun flower, but not on any flower of a different colour.

It is well worth while that this appearance should be attended to in this country.

Mr. Medicus[61] at Manheim has several singular ideas concerning the production of mushrooms; he denies the existence of any Seeds in these. It is affirmed by the same Gentleman that Mushrooms spring only from Vegetable substance; this I think may be very fairly controverted.

Further, he affirms that they do not grow when the vegetable is in a state of putrefaction, but when in the first stage of decomposition; but this is nothing else than what others would call the first stage of putrefaction.

He ascribes the growth of mushrooms to an elastic and attractive force in the parts of the decayed vegetable; in consequence of which he calls it a vegetable Crystallization. This is a new attempt to establish the equivocal generation of plants; he has been followed by his countryman Mr. Necker but without any foundation.

Mr. Deride[62] proposes another example of Crystallization of organic bodies. It is a rare production but little known, but with which I have been long acquainted. It is described by Weber[63] by the name of *Lichen radiciformis*. It is found in large quantities in mines upon the decayed wood; it is what the miners here call Trash. It is affirmed by Dr. D. Rutherford[64] that it is not a plant and that it is found nowhere but in mines but this may be easily controverted, for I have found it growing upon the surface of the earth but never unless between the wood and bark of decayed trees. It being one of those species called *Lucifugi*.

The last article I mentioned was the doctrine attempted to be established by Monsr. le Riney[65] which he calls Vegetable Crystallization; he carries this doctrine farther, he likewise found in mines other Cryptaceous [cryptogamic] plants, agarics etc. All these are considered by this Gentleman as mere effects of Vegetable substances in what he calls the first stage of decomposition. We are however perfectly acquainted with the fruit and seeds of several of these plants.

It used long ago to be a favorite subject with authors the vegetation of

[61] Friedrich C. Medicus (1736–1808).

[62] Unidentified.

[63] Georg H. Weber (1752–1828), *Spicilegium florae germanicae* (Göttingen: 1778). *L. radiciformis* is not a lichen.

[64] Daniel Rutherford, professor of botany at the University of Edinburgh.

[65] Possibly René Antoine Ferchault de Réaumur (1683–1757).

Stones, but now the tables are turned and a great number of vegetables are supposed to grow by Crystallization. Both these doctrines are equally ill founded.

A Physician settled at Madras; Dr. Anderson, whose attention does him great credit has pointed his researches to real use. He has sent from India to the Island of St. Helena seeds of several useful plants to be from thence transplanted to the West Indies. Among these is the Tree which yields the true gum Arabic; the ———— the Juice of which is the *Terra Japonica,* and the Gossipium or silk cotton tree.

Mr. S———— in Germany gives an account of a plant which is called in England the cerean[?] silk weed. It is supposed to be useful in the manufacture of Cloth.

A treatise has been lately published at London on a new species of American Grass, the *Agrostis cornucopiæ,* from South Carolina; the seed of this plant has been brought to London and sold at two Guineas the English Gallon; its qualities however are rather problematical. We have several of the same Genus in our own country; besides little is to be expected from it as it is one of those plants that grow in woods, and few of these are coveted by cattle; and in all the trials of the seed that I have heard of the grass has never appeared.

In the transactions of the society of Arts at London published last October, there are several instances that show how much natural history may be applied to several arts.

The Agricultural Society of Paris has published a volume in which there are several instances of the same thing.

In the former we see several cases by Lord Fife,[66] of roots that are commonly cultivated. 100 square yards produced 77 stone of the common Turnip, the same portion produced of Carrots 73 stone. In the same Volume we have several remarks concerning the Larch tree. The more we are acquainted with it the more we may be thankful we have got possession of it in this country. There is one which is the tallest in Britain at Blair in Atholе; it was planted in the year 1764 and contains 130 feet of square timber and would sell for six Guineas; it is within a few feet of a 100. The publick spirited Bishop of Landoff has planted 48000 larch trees.

In the Commentaries of Göttingen, Professor Murray[67] has discovered the real plant that affords the true Gamboge; the plant termed by Linnæus

---

[66] James Duff, second Earl of Fife, F.S.A., F.R.S.

[67] Johann A. Murray (1740–91), professor of medicine and botany at Göttingen and a student of Linnaeus.

the *Gambogia gutta* was supposed to produce this resin; but Dr. Murray has found that it is a different plant, he called the ———.

Here I would observe that Authors have been very far from being sufficiently correct in Generic names. Linnæus laid down rules for this purpose and these have been assented to; but they seem now to be lost sight of. For example, that name ——— of Professor Murray is improper, because the three kingdoms of nature ought not to be confounded together; and besides all names ought to be avoided where there is too great a similarity, either in the orthography or pronunciations. Another rule never to be transgressed is that one Generic name ought never to appear in the form of a derivative from another. All barbarous names ought to be rejected.[68]

From all these we see that the names which have been lately given to several plants were not constructed by Linnæus.

We come now to the discoveries in

ZOOLOGY

There is a magnificent work lately published which has not yet reached this place, the *Zoologia Danica* by Muller.[69]

In the last volume of the Göttingen Commentaries there is a curious paper on the living powers of the Blood in animals; "That there is a principle of quality in the Blood capable of producing effects which are not to be explained by the mere physical properties of matter." The first who made this observation was Fontana. This doctrine has likewise in some measure been adopted by Mr. Hunter of London, but is opposed by Bloominteach.[70] The violent motion of the blood which has been demonstrated is not, says he, owing to any thing in the Blood itself, but altogether to the living power of the heart; inject a solution of Isinglass into a heart newly opened and you will perceive the same effect as from Blood. The generation of new vessels is adduced by Mr. Hunter as an argument in favour of the living power of the blood; but according to Bloominteach this does not take place from the blood but only from the Lymph.

Of late there have been some curious dissenssions [sic] concerning the natural history of Ambergrease; this substance has at different times been referred to all the three kingdoms of nature.

---

[68] The problems connected with the formation of new names are still with us. It is interesting to note that such difficulties in classification were recognized at an early date.

[69] O. F. Müller, naturalist.

[70] Perhaps Johann F. Blumenbach, student of comparative anatomy and physiology at Göttingen.

Dr. [Scheuchzer?] supposes it an animal substance; he found it in the Intestines of the *Physalis macrocephalus* of Linnæus, and in those whales which abound upon the coast of America and have laid the foundation for the southern fisheries.

This animal lives upon the *Sepia artipodia* and these last are remarkable for a peculiar hard horny bill and this is often found in the substance of ambergrease. The well known analysis however of ambergrease, militates rather against this opinion. It appears to be composed of an Oil and an Acid Salt in a crystallized form. We could not expect this upon the distillation of an animal substance, the acid found in human *Calculi* will hardly invalidate this objection. It is known very well that considerable numbers of that species have been killed in the European seas, yet in these there was never found the least appearance of Ambergrease.

In the last Volume of the Philosophical Transactions [Royal Society of London] there is a meritorious paper by Dr. Grey[71] upon Amphibia. Linnæus defines the Class of Amphibia in general to possess only a single heart and no doubt in this he followed the opinion of Boerhaave;[72] but in many of that class the heart is double with an immediate communication between the two cavities.

Dr. Grey is at great pains to detect the difference between the venomous and harmless serpents, and he does this either from external structure or the structure of the teeth. The venomous serpents are remarkable for the breadth of the head, the sharpness of their Tail and last of all there is a Chain of Scales on a long protuberant line running down their middle like the keel of a boat; this however I think is liable to objection. With respect to the Teeth he attempts to determine concerning the fang. The most material distinction however is this; according as their teeth or fangs are solid or tubular.

In the teeth of a rattle snake which I now show you, the ven[o]mous nature of the animal is evident from its tubular form; the orifice is placed upon the convex side of the teeth.

Dr. Grey further proceeds to divide serpents generically.

Having now finished the consideration of the discoveries and improvements in natural history since last year, I proceed next to the

[71] Edward W. Gray, "Observation on the Class of Animals Called Amphibia . . ." *Philos. Trans.* 79 (1789): 21–36.

[72] Hermannus Boerhaave (1668–1738), professor of medicine at Leyden.

DEFINITION *of* NATURAL HISTORY

and I would define it to be:

The arrangement, description and history of the works of creation contained in the terraqueous globe.

The history of the Heavens is the subject of another Science that vizt. of Astronomy.

Physics formerly called Physiology and now natural philosophy is the knowledge of the properties of the phenomena of nature, of the powers of matter and motion and their causes and effects. Natural History having for its object the sublunary works of the creation.

To avoid being bewildered amongst such variety, it is necessary to proceed with all the precision we are able and first to mark all the different branches of the science. It consists of the six following:

1. Meteorology
2. Hydrography $\Big\}$ Contain the natural history of the globe in general as marked
3. Geology[73] by Hippocrates.

4. Mineralogy
5. Botany $\Big\}$ Contain the natural history of the fossil; vegetable and animal kingdoms.
6. Zoology

These compehend the whole of natural history, and all of them are comprehended under it.

In the three first consists the natural history of the terraqueous globe in general, i.e. of the Atmosphere, of the waters and of the earth. This triple division was anciently marked by Aristotle in his book, *de aere, aquis et locis.*

The three last comprehend the numerous individuals which compose the three kingdoms of nature, the Fossil, the Vegetable and Animal.

To give a more exact Idea, I shall take a nearer view of the six branches and draw the outlines of each, and point out more particularly the subject of our intentions.

Beginning with

METEOROLOGY

This is to be understood as the natural history of the Atmosphere, of its general phenomena and those appearances which occur in the various

---

[73] Although the word geology had appeared before Walker's time, this is the first use of the word, so far as I know, to designate a specific portion of natural history covered in a series of lectures. All of Walker's "mineralogy" and most of his "hydrography" are now included in the science of geology.

countries and Climates. Of the Barometer; the vicissitudes of the Atmosphere which correspond to its rise and fall, the cause of its increase and diminution of weight, the mensuration of heights by means of that Instrument; an Experiment which has an extended influence in the history of all the three kingdoms, "whether it can give an indication of the changes of the weather?"

Next we come to the variations of the Atmosphere with regard to heat and Cold as ascertained by the Thermometer, the formation of Ice, the production of Snow and Hail, their various configurations, and where they differ from Ice in their congelation; the formation of hoar Frost, the reason of its appearance in summer. Here we shall attend to the point of congelation in the Atmosphere; where water is converted into Snow, and to the point of perpetual congelation where of course it must be perennial; the extraordinary appearances that attend the highest degree of heat and cold both in this climate and others; the theories of Freezing; the expansion of fluids; the contraction of solids; the effects of Frost on Animal and Vegetable life; the gradual increase of Cold as we ascend in the Atmosphere. I shall consider of this principally to show the elevation of places by the Thermometer; the degree of heat and cold in the different climates of the globe; the range of the Thermometer in different places and the diversities of heat and cold in countries under the same parallel; the different states of the air will be noticed as discoverable by the Hygrometer.

We shall enquire into the Theory of evaporation and the suspension of the Clouds and the attraction between the clouds and Mountains; and consider the nature of the aqueous meteors, Rain, Dew and Fog, particularly the comparative depth of Rain in different places through the year; why the Mountain countries are the most rainy; and those countries that are most destitute of rain; the opinion concerning the formation of Dew; whether it ascends or descends, its attraction to sharp points, and its most common appearance in spring and autumn; its noxious effects on the human body. The cause of Fogs, whether marine, terrestrial or Aerial; likewise the different heights of the Atmosphere in different places; the nature and origin of the land and sea breeze appearances which prevail in the Tropics; of the trade winds; of the winds in the Indian Sea; of the storm which proceeds from the Bulls eye; of Whirlwinds and Tornadoes; and of general and local hurricanes; to conclude with an account of the different qualities of winds and their influence on mankind. The history of luminous meteors; the Rainbow; the oval figure sometimes assumed by the Sun. The history of Lightning; of the *Aurora Borealis,* and other Electrical appearances, the *Ignis fatuus, Draco volens, Castor and Pollux* etc.

We shall likewise touch upon the doctrine of sound as from natural observation; the progression of Sound thru the Atmosphere, and its various modulations by means of Echoes.

Of the various and important effects on the Vegetable and Animal Kingdoms.

The variation of Climates produced by the situation of places with respect to Land and Sea, and the heat of the air; the difference between the north and south in different places under the same latitude. The rise and progress of Epidemical diseases conveyed by the air, being obnoxious to plants and Animals. The nature of blights and their bad effects on many useful crops may here be considered, and the false and ridiculous causes to which they have been ascribed.

The succession of seasons; whether they move in a Cycle of any fix'd number of years; their peculiar properties, their variations in different countries, and in the same place naturally afford some speculations.

This part of our subject will be concluded with an account of those appearances which indicate an alteration of the weather, with observations on the ——— of it and their Causes; and a new form of a Diary of the weather.

The IId branch of [natural history] is

HYDROGRAPHY

or the natural history of the waters of the Globe.

We begin with considering the different sorts of waters, and their peculiar properties, as Rain, Spring, River, Snow, Dew, Lake, Sea, Pit, Waters, etc.

The nature of these waters which produces hurtful effects on the health. Likewise the singular effects of oil in levelling the surface of water. The history of the ocean, its Shores, its harbours, its various depths and the nature of the submarine regions; the origin of its saltness undetermined; whether it is naturally salt or acquires its salt in the course of ages; whether its degree of Saltness is fix'd and stationary or if there is reason to show that it may increase or diminish.

The different colours of the sea will be observed, and its luminous appearance, the cause of which is still unknown, nor has even the Phenomenon been fully described.

The history and Theories of the Tides, the variety of appearances they assume in different places, and particularly the origin, progress and direction of the tides in the British Seas.

Currents etc. in some parts of the Globe also deserve our particular attention.

Likewise the Water Spouts frequent in the warmer climates and especially that kind called Typhon in the sea of China.

The peculiar nature and properties of the Inland seas, such as the Baltic, the Mediterranean, the Euxine, the Red Sea etc.

The streams in the Straits of Gibraltar. Likewise the gradual recesses of the Sea from the Earth, and its encroachments on the land by accidental Inundation.

The different sorts of Springs; whether perennial, intermitting, periodical, or temporary. A view given of their heats discoverable by the Thermometer.

Next we proceed to the rise and course of rivers, the formation of their beds, the quantity of their waters and their accidental and periodical inundations.

The formation of Lakes, their depths; the quantity of water they receive and discharge; their evaporation; and a particular description of those which are remarkable in these countries for resisting the frost.

The degrees of frost. The circulation of the Atmosphere to the Earth; of the Earth's to the ocean, and of both to the higher regions.

The IIId branch of which we shall treat is

Geology

or

The Natural History of the Earth

Here we will view the fabric of the Globe as distributed into Land and Water. The general description of the Earth with respect to Mountains, and particularly the westerly elevation observable on the Continents, promontories and Islands.

The general shape of the Earth in different climates.

The form, course and similarity of Shape observable in all contiguous Mountains, and those of the same chain. The general direction of all Mountains from South to North. The direction of [continents] regulated by the direction of the Mountains.

All Mountains are distinguished into two Classes, primary and secondary, from their being formed at different times.[74]

These distinctions made by Fossilists[75] require to be fully explained if we would have a proper idea of the structure of the earth; the two kinds of

---

[74] It is significant that Walker associates the origin of the two classes of mountains with time. The idea was not original with him but goes back to Lehmann (1756) and Arduino (1760).

[75] I.e., geologists—a common usage before the 1790's.

mountains are to be accurately compared with respect to position, their Latitude; their inclination to the horizon, their strata, and mineral contents.

The disposition and formation of the various Islands; their extension; the direction of their adjacent continents; and how they differ from continental lands.

We shall animadvert on some opinions concerning some important alterations of the globe, of which there are now many vestiges. Such as the destruction of the great Island Atlanta; the separation of Britain from France by the British sea; the [separation] of Europe from Africa at the Straits of Gibraltar; etc.

The disposition and variety of the Strata of the Earth will next become the subject of our particular investigation; as they are distinguished into primary, secondary and accidental. We shall remark their depth, and inclination to the Horizon; the correspondence between the Strata and the surface, and whether the strata are disposed according to their respective specific gravities.[76]

We shall examine the origin and progress of the staple of the Earth; the Vegetable mould; the large masses of stone that are everywhere spread over the surface of the earth; the origin and nature of turf and peat, with some observations on the peat mosses of our own country (a production that is peculiar to the colder climates) and other factitious strata.

We shall next consider the nature of quick sands on the shores of the sea and of rivers.

The nature and origin of sea sand, its baneful invasion on some countries and its rise as a barrier against the encroachments of the sea in others.

We shall next give an account of the veins observable in the earth; its Causeways; its Caverns and Grottos, with the most remarkable subterraneous discoveries.

We shall then give a description of the several sorts of damps generated in mines, and the different Mineral steams issuing at the surface of the earth. Then we shall make some animadversions on central fire and subterraneous heat, and these being duly considered, will prepare us for entering on the History of Volcanos and Earthquakes. First we shall take a view of the situation of Volcanoes; the cause of their inflammations; their projectile forces, their extinction; their bursting forth *de novo* and the relief they give from earthquakes; with some account of their mineral productions.

As to Earthquakes; we shall examine their progress, their concomitant and

[76] Many of the leading philosophers of the eighteenth century believed that sediments had been precipitated from a universal sea according to their specific gravity. Walker later argues against this concept.

consequential phenomena, which are the only means by which we can discover their cause.

Lastly, we shall make some observations on the soil of different places with respect to wholesomeness.

These conclude our remarks on the three first branches of our course. The history of the atmosphere, of the water and of the earth, or what was treated of by Aristotle under the head of *Aere, Aquis* and *Locis.*

We then enter upon what Philosophers call the

IMPERIUM NATURÆ
the
EMPIRE *of* NATURE

which is divided into three kingdoms, namely the Fossil, Vegetable and Animal.

Lord Bacons great desideratum, a catalogue of the works of nature, is here to be considered. One great object of natural history is to arrange the bodies of these kingdoms in such a manner that each individual may be properly distinguished and certainly known; but their number being immense a regular system for this purpose is necessary and indispensable.

In the first place the rule of method must be fully and carefully explained, for without these it is impossible to become a scientific naturalist.

The utility of the Aristotelian method of division into Genus and Species will be considered.

The modern one into Classes, Orders, Genera and Species tho' arbitrary deserves much commendation, and is to be adopted as the most comprehensive and commodious.

The natural, essential, and factitious characters of Genera will require full explanation.

The number of Species in the three kingdoms are to be attended to and observations will then occur in what is termed the Scale of beings, of which there are so many component parts.

There is undoubtedly a continued chain in nature from its lowest subject up to the human species, and which it is to be supposed proceeds from him to his maker, all being linked as in the moral world by the most beautiful and regular gradation; for nothing is more certain than the maxim, *"Natura nunquam fit saltus."*[77]

[77] "A skip is never made by nature." This paragraph shows that Walker had an under-

The manner of philosophysing [sic] of Natural history, and the method of proceeding from the *majus* to the *minus nota* by an exact and scrupulous Synthesis must be pointed out.

But tho Systems may be constructed in this manner, they must be used and taught by Analysis.

We shall consider the knowledge of Fossils, plants and Animals according to what is called a natural and artificial or mix'd system.

These promised, we will then proceed to survey the

Fossil Kingdom

constituting the science of *Mineralogy*.

Here we will begin with the history of the science and endeavour to trace its rise and progress and to describe its present state; from which it will appear that of all the branches of natural history, Mineralogy is at present the most imperfect.

The best systems do most confessedly labour under many very particular defects; nor is there any one so well calculated for our purpose as could be wished.

I shall therefore be induced to offer to your consideration a system of Mineralogy or rather the rudiments of a System which from a prospect of more time and opportunity I shall hope to render worthy your attention.

Here we may view in a cursory manner that curious but abstruse question. Whether all the Fossils in the Globe have been formed from one simple and original earth? And it need be no objection to us in the prosecution of this enquiry that no satisfactory opinion has been received on it.

There are however earths more obvious and with which it is of more moment to be acquainted; which are the most simple that we know and serve as the basis of all Fossils[;] these are called primitive earths. These are the foundation of the science of Mineralogy.

All Fossils are comprehended in three divisions:

    I. Earths
   II. Stones
IIId. Minerals

Minerals are again separated into three subordinate divisions, vizt.:

    1. Salt
    2. Sulphur
    3. Metals

---

standing of the gradational nature of life forms with man at the top of the pyramid of life. In *Institutes*, p. 71, he says, "Varieties, though they change by seeds, continue strongly by evolution."

and the last being accounted the most important have given the name of Mineralogy to the science.

These divisions comprehend the natural history of Fossils; but to these may be added the extraneous fossils[78] which are either animal or vegetable petrifactions.

Under the general delin[e]ation of Fossils we shall mark the distinctions and define the terms peculiar to their situation, figure, substance, consistence, and structure, their parts and qualities and their origin, decomposition and Regeneration. After the terms of the science and the rules of arrangement are explained; We will proceed to define and describe the Classes, Orders and Genera, and in some degree to characterize and distinguish the species and varieties of Fossils.

It must be acknowledged that this is a matter of the greatest difficulty, and impossible to be executed, in the present state of the science, as we could wish.

But all the difficulty arises from the nature of the subject; for as fossils proceed from no seminal principle, their formation is always obscure and the species always difficult to determine being circumscribed by no certain boundaries but running into one another like the shades of a picture, so that it is almost impossible to say where the one begins or the other ends.

But however tho' we cannot every where distinguish these clear and certain marks, we must do it in the best manner we can.

To that of the Fossil Kingdom, may be very properly subjoined an arrangement of Mineral waters and an investigation of their contents origin and Medical virtues.

Here we will make some particular enquiries concerning the Fossil kingdom, as ———.

A general view of the distribution of the Fossils of the Globe; when I shall suggest the form of a Mineralogical Journal[79] to be kept by gentlemen particularly those who trav[e]l, and which will in my opinion facilitate the improvement of this science more than anything of the kind that has been attempted.

After this we will come to the theory of petrifaction and enquire whether this is performed by Induration, Incrustation or Insertion.[80]

Then we will attempt a history of petrifying waters, with an examination

---

[78] Fossils in the modern sense.

[79] Walker's own *Mineralogical Journals* were written between 1759 and 1779. Jameson in 1799 and 1800 followed the same practice.

[80] Later in his lectures Walker enlarges upon the methods of petrification (fossilization). This is probably the first attempt to understand all the methods by which organic material is fossilized.

of the opinion concerning the vegetation of Stones and minerals; the formation of Crystalyzed stones.

The history of Gems, their colour, figure, internal structure and how far they may be imitated by art.

We will take a view of the production of Amber and Ambergris; of the Turquoise stone; of the columnar Basaltes; of Shell Marle and other substances of doubtful origin.

Then we shall make an enquiry into the disposition of the metals in the bowels of the Earth; into their direction, dimension and contents. Of metallic veins with a description of mines. In Germany and Sweden the study of Mineralogy is successfully cultivated and liberally encouraged under the direction of government; for the interest of the state is concerned in the improvement of mines.

In our country such an improvement would certainly not be impolitic; for if Scotland has not made a figure in this article it must certainly be ascribed to want of industry, want of Skill or want of Encouragement; for no European kingdom promises so much or has produced so little.

We shall also enquire into the Mineralization of ores; the causes of their change and regeneration in those places which had formerly been deprived of them; the imitation of metallic ores by art, and many of them can be imitated this way, for in most of these tho' their principles be known, they cannot be imitated, from our ignorance of the secret operations of nature, and their slow growth in the bowels of the earth.

It is an established principle that all metals are coeval with the Globe; for no earth has ever yet produced them from any substance in nature but what was already metallic; but being liable to decomposition they pass thro' various changes in the Earth, before they arrive at the state in which we find them.[81]

In examining the species of Fossils many curious and particular enquiries will occur; some of them of a philosophical nature, such as the singular property of the Canary Stone for the filtration of water; the magnetism of the ores of Iron; the double refraction of the Iceland crystal; the phosphorescent properties of the Bolognian stone, and other enquiries connected with manufactures and the arts, such as the properties of different clays in the making of bricks, and porcelain etc. The properties of Fullers earth and the detergent qualities of Soap.

---

[81] Walker is telling his students that metals do not occur in their original state but have undergone various changes during their long history before reaching the state in which we find them. He considers the source material of metals coeval with the earth, but realizes that it may be altered, transported, and redeposited.

We will treat a little of the Mineral earths serving as colours in painting; of Tripoli and the species of rotten stones used in cutting and polishing of metals; and of the different sorts of limestones for building and manure; of the ancient and modern marbles; the Jaspers; the Porphyries; the g[ra]nites etc. Of the calcination of Gypseous fossils, of which the plaister of Paris is made. Of the fabrication of paper and cloth from the stone called asbestos. Of hones and other stones made use of for sharpening knives. Of the different stones made use of in building and of Slates used in covering houses.

Of the use of the Vitriols of metal in the arts. The symptoms and method of discovering Coal and Lime stone. The causes of the blowing of freestone and others in Architecture. Of the causes of dampness and the admission of water into houses from the nature of the materials and stones of which they are built.

Of the operation of Lime as a cement; and with some account of the mineral poisons.

Having taken this extensive view of Geology and Mineralogy we will then enter upon the curious and much agitated subject

The Theory of the Earth

Here we will endeavour to trace the changes which our globe has undergone.

The method of enquiry which all our ingenious Theorists of the Earth have pursued is certainly erroneous. They first form an hypothesis to solve the phenomena, but in fact the Phenomena are always used as a prop to the Hypothesis.

Instead therefore of attempting to cut the gordian knot by Hypothetical analysis, we shall follow the synthetic method of enquiry and content ourselves with endeavouring to establish facts rather than attempt solutions and to try by experiments how far that method may lead us thro' the mazes of this subject.

Having proceeded thus far on the Inanimate parts of nature and considered the three great masses of the subject Air, Water and Earth; we will then proceed to view the

Vegetable and Animal Creation

where nature presents us with beings endowed with Life and a power of propagating their kinds.

But there are some enquiries common to both of these kingdoms which require to be discussed before we begin to treat particularly of either. Such as———. The Species and varieties of plants and Animals with the history of their generation.

The doctrine of Equivocal generation, the multiplied if not as some contend the Infinite envelopment of plants and living creatures in their seed and eggs.

The propagation both of plants and animals otherwise than by seeds. The production of Mules both in vegetables and Animals; and the noted infertility of these hybrid productions a most remarkable and necessary appointment in nature.

The doctrine of Equivocal generation is here to be considered and rejected upon review; and that which is called universal generation explained and established; and from the consideration of these the following consequences will appear natural and probable, vizt. "that no species of plant or animal is ever changed into another" and that no species has ever been lost since the creation of the world, or any new one formed.[82]

We now proceed to the particular consideration of the Vegetable Kingdom, that is

BOTANY

This consists in the arrangement; denomination, description, and demonstration of plants which makes a considerable part of their natural history; but this it is unnecessary to handle here, it being so well taught in another place.

It is the business of a Botanist to methodize and distinguish plants.

It is the part of a Naturalist to be acquainted with their properties.

It is the task of a Botanist to discover unknown plants with a view to their future usefulness.

It is the business of a Naturalist to discover useful qualities in those that are already known.

We shall pass therefore over the mere nomenclature of plants and confine ourselves to the consideration of their general nature, their powers and properties. In fine it is not a botanical but a philosophical history of Vegetables and Vegetation that we propose.

We shall therefore begin with taking a general view of the structure and organic parts of vegetables.

[82] Although the doctrine of equivocal generation is rejected, Walker is short of the truth in the final statement concerning species. We must remember, however, that the full significance of fossils had not yet been grasped. This is, indeed, a curious statement in the light of other comments on fossils, species, and reproduction, quite contrary to other of his statements on evolution.

We shall examine the different kinds of motion of vegetable fluids, the course and manner of the ascent of the Sap; the laws by which its motion is directed; the doctrine respecting the circulation of the Sap; and whether such circulation exists. The generation of seeds the nutrition of plants; their perspiration and absorption by the Leaves, and other matters relative to the vegetable kingdom.

The different parts of plants are divided into two Classes.

1st. Those which serve the purposes of Nutrition.

2d. Those that are essential to fructification.

Here we are to examine the motion of the different parts of Vegetables, especially of roots and stems; and we shall attempt a solution of that curious problem. The descent of Roots into the Earth and the ascent of Stems into the air, with the surprizing movement of stems to preserve or regain their perpendicular direction, and the motion of branches to preserve a position parallel to the surface of the Earth.

We shall next observe the positions and motions of the Leaves and consider that universal distinction of an upper and under surface, the cause of this appearance and its important effects in the Vegetable kingdom.

We shall then suggest the uses of the different parts of plants and the different offices they seem intended to answer.

We shall take notice of the general disposition of plants over the Globe and the methods of their [dissemination] and disposition and that coincidence which is observable in all plants, with the climates in which they are found. This will lead us to trace the effects of soil and climate, in fixing the situations of plants in particular places called the *Stationes plantarum,* and here we will naturally remark the similar properties of plants that grow in similar situations.

We shall then consider what is called the *elevatio plantarum*—i.e. why all plants affect to grow in a particular region of the Atmosphere; some on a level with the sea and some at a greater or less distance from the Level.[83]

All plants are more or less sensible of the vicissitudes of the weather; heat and cold; Dryness and moisture. Many plants upon the suns recess change the position of their Leaves and maintain a different position in the day from what they do in the night—this is called the *somnus plantarum.*

There is also a similar appearance in Vegetables called the *vigilia florum,* which is the opening and closing of their flowers according to the state of the

---

[83] *Elevatio plantarum* is today a part of plant ecology. Ecology did not become a separate discipline until the nineteenth century. For a history of ecology see Richard Brewer, *A Brief History of Ecology, Part 1, Pre-Nineteenth Century to 1919* (Occasional Papers of the C. C. Adams Center for Ecological Studies, No. 1 [Western Michigan University, 1960]).

weather and in some plants the flowers even open and shut at fix'd and particular hours.

There is yet another phenomenon to be illustrated, which more than any thing exhibits the extreme sensibility of Vegetables, vizt. the influence of Light on plants. This it is which directs all their motions and it is this which cloaths them with that beautiful verdure so grateful to the Sight.

Afterwards we shall treat of what is called the *gemmatio* and *vernatio Arborum,* or the Budding and Foliation of Trees; and shall propose to you the manner of keeping [an arboretum?], which we will have occasion to recommend, as not only conducing to the pleasure of such as are disposed to agrestic studies, but which will be attended with profit to such as are inclined to a rural life.

Here it will be proper to make a few observations on the habit of plants; then will follow some remarks on the colour of Vegetables in their natural state and the effects produced on that colour by mixture.

We shall consider the different odours and tastes of vegetables and enquire how far these point out their mechanical or medical properties, and tables will be given of these sensible qualities.

The medical properties of vegetables [are] being treated of elsewhere. We shall take a survey of their numerous and important uses in life and enquire how far their habit and external qualities are capable of indicating their medical virtues.

But our principal attention will be turned to their economical qualities and utility in the Arts.

With this view we shall most carefully remark the principal parts of the most useful plants, as barks; roots, fruits, seeds, and the spontaneous vegetable exudations as manna, honey etc.[,] with the different kinds of Gums or resins, the most useful parts of plants.

We shall particularly remark the esculent plants, especially those of our own country tho' not at present in general use yet which may be used in diet, not only with safety but with advantage. We shall point out the vegetable poisons and particularly the deleterious plants of our own country.

The enquiries respecting the vegetable kingdom in general will be concluded with a specific account of such plants as are possessed of any rare or remarkable properties or are useful or noxious to Mankind.

As a course of Lectures of this kind should be of public as well as private utility every opportunity should be embraced that can in any way be applied to the advantage and improvement of the useful arts, and such an opportunity occurs here.

The observations to be made and the principles to be established from the several subjects now enumerated may be of great use in the several arts dependent on Georgics.[84]

These consist of two parts:

1st. The cultivation of Plants and

2d. The management of domestic Animals

The latter of these will come in more properly under our view when we come to treat of the Animal kingdom. The former may be the subject of our consideration here.

Of Georgics the first and most important branch is Agriculture which tho' only an art in its practice may be justly considered a science in its Theory and principles and is intimately connected with the philosophical history of plants. Of this however we cannot propose to treat in its full extent, but only so far as it depends on the principles of vegetation.

In this view we shall consider the nutritious matter of vegetables and what is called the Pabulum of plants.

We shall treat of the nature of soils in general and of those in Scotland in particular, with their particular properties, names and distinguishing marks, and the particular plants each of them is fitted to rear.

Of the operation of Natural and Artificial manures especially of Quicklime. Of the effects of Tillage. Of the difference between horse and other methods of Husbandry.

Of the structure of Roots. Of the change of species and rotation of Crops. Of the comparative merit of the different grains and other profitable crops. Of their different effects on the soil and of the nature of propagation and destruction of weeds. Of pasture and meadow lands. Of the culture of Artificial grounds.

Of the discovery of some grounds and other plants not now in use, but which are fitt to be tried as Green or Dry forage.

Of the reclaiming of wild land, and to these we shall add a review of those obstacles which obstruct the improvements of Agriculture in Scotland.

There is another Agrestic art which is nearly allied to Husbandry as depending on the Natural history of plants, vizt.,

Gardening

Here we will begin with observing the effects of cultivation on plants which in consequence of the variation of climate and the course of ages exhibit to us that vast variety and considerable improvement which nature may be brought to when assisted by art.

[84] Rural affairs.

We shall take notice of the luxuriance and singular appearance of Plants which recede from nature in their leaves, flowers and fruits. Of the Striped, blotched and Crisped Leaves. Of the double, proliferous and other monstrous Flowers, and, of the vast variety of excellent fruits from one original.

Of the naturalization of Plants. Of the detriment they sustain by change of climate, Heat and Cold etc. Of the means to be used for their preservation, and of enabling them to endure a remote climate. This deserves the particular attention of such who like us live in a Cold climate where most of the valuable and useful plants we possess have been brought from the warmer climates, and from whence we may hope for the acquisition of many more.

We will take notice of the diseases of Cultivated Trees and other Garden plants. Of the origin and History of Mildews with their baneful effects and remedies.

Of the means of preventing the destruction of plants by Caterpillars, with remarks on the erroneous tho' common opinion that the sea air is prejudicial to vegetation.

We shall then come to the different operations of Gardening for propagating and improving plants, such as sloping, laying, pruning, grafting and inarching. Taking particular notice of the summer and autumnal shoots of Trees, a circumstance of importance tho' but little observed and hitherto unexplain'd.

We shall then consider the different stiles of Gardens and Gardening; the Kitchen and flower Garden; the management of the fruit Garden; the construction of Fruit walls and the different sorts of shelter for fruit Trees; the Orchard, the shrub[b]ery, and the Botanic Garden; the Green house, the dry and bark stoves and other conservatories for tender plants, and the propagation of valuable fruits.

We shall then bestow some observations on that higher species of Gardening the laying out of pleasure ground, an Art that not only requires an extensive knowledge of natural history but of the human heart.

We shall then come to the translation of plants, i.e. the Transportation from their native places and their establishment in a different country and climate; a process in which not only Gardening and agriculture but the conveniency and wealth of Nations are deeply interested.

Lastly, we shall come to the art of planting, and this requires particular attention in a Country which like this is advancing in improvement, for here the natural woods must be destroy'd and give way to fields and pastures.

We shall attend to the rules of transplanting, pruning and topping of Forest trees and give a catalogue of such as are fit for growing in the open air of our own country and examine their qualities and comparative merit.

We shall observe the distinction between Evergreen and deciduous Trees and the Cause of this important difference. The formation of woods by sowing or planting. The cause of the destruction of Natural Woods and the means of restoring them.

The designing of Clumps and Groves and the formation of Underwood. Of Grouping of Trees by their sizes, shapes and the Colour of the Foliage and of the best method of transplanting full grown trees.

Of the humble tho useful tribe of Arboreous plants, the Willows, the Oziers etc. of which the wicker work is made.

Of the different sorts of Bark for the use of Tanners.

Of the progress of Arboreous plants in general.

Of the disposition of their Roots and Stems.

Of their Longevity and the nature of the Ligneous circles which fix the Chronology of Trees.

Of the difference between the Red and White wood and the use of the Blea in timber.

Of the provision in the trunks of Trees for their annual increase and the most proper season for felling Forest timber. Of the comparative value of different timbers, and Lastly We shall attempt a history of the rise and progress of plantations in Scotland with remarks on its present state, and the means of its further advancement.

We now come to the sixth and last general branch of natural history vizt.

ZOOLOGY

OR THE

HISTORY OF THE ANIMAL KINGDOM

Here I shall begin with remarking the arrangement of Animals over the Earth in general where we will have occasion to view a striking ordination of divine wisdom in adopting the different Animals to the different Climates in which they are placed.[85]

We will examine the senses of Animals, their means of subsistence and the dependence of the several tribes on one another, like different links of one chain. Here the wonderful contrivance of nature for the preservation of the Individual as well as of the Species will be a very pleasant subject for speculation.

The fertility of Animals seems capable of being determined by established laws, but that the number and proportion of the individuals in each Species

---

[85] This discipline today is known as animal ecology.

should be so exactly preserved seems to be beyond the reach of any general law. The various degrees of Fertility in different Animals so exactly adapted to their circumstances is a necessary and indispensable provision of nature.

We shall likewise remark the inferiority in Fertility of some animals when restrained from their natural manner of life, and the superior degree of it in others when in a state of domestication.

When we enter on the methodical arrangement of the Animal Kingdom, the system of the celebrated Linnæus will be our guide in preference to all others. It is a system the best adapted to use and the least discordant to nature, it is the most accurate in its arrangement, and the knowledge of its terms and characters is indispensably necessary to every one that would become proficients in the study of natural history.

In this system all the animals on the globe are comprehended in six general Classes, each of which must be handled separately and in detail.

The characters of the several Orders and Genera must be explain'd and an account givn of the most important and most remarkable Species in each.

In this System the first Class is entitled

Mammalia

This contains all the Terrestrial Quadrupeds, and which was formerly used to be termed Quadrupeda by Naturalists as indeed it was by Linnæus himself in the first edition of his System; but he afterwards adopted into this Class all the Animals of the whale kind because they differ from fishes and correspond with this Class of Animals in many circumstances, such as in breathing thro' Lungs; having warm blood, a double heart with two Auricles and two Ventricles, etc. He therefore dropt the term Quadrupeda and assumed that of Mammalia.

There belongs to this Class a great variety of Animals, some remarkable for strength, size and ferocity; others for the greatest sagacity and peculiarity of manners; some as being most useful and advantageous, others most formidable and pernicious to Mankind. It will therefore deserve our particular notice.

Here we shall treat of the ruminating Animals with attention, and of the [problems?] which relate to the management of domestic Quadrupeds, and endeavour to note the effects of domestication in the several species.

We shall then consider the means of improving the breed of Cattle, and take notice of such as may be ben[e]ficial to mankind by being rendered domestic.

The second Class in the system of Linnæus is that of the Birds whose natural history constitutes what is called

Ornithology

Here the External parts and characters of Birds must be described and the terms of art relating to them explain'd.

Then we shall attend to the more obvious division of them from their situation and manner of life, and that founded on their external appearance affording certain marks for distinguishing them.

We shall observe the methods of their propagation, incubation, and the different manners of rearing their young. Of the peculiarities in the structure of their nests, of their food, of their longevity; of their different manners as inhabiting land or water as solitary or gregarious and of their usefulness in contributing to the subsistence of mankind.

In surveying the orders, genera and species of this Class, we shall pay particular attention to the migration of Birds. We shall enquire into the causes and endeavour to determine the countries to and from which birds of passage remove at certain periods.

Here it may not be amiss to take a little notice of the training of Birds of prey for sport. Of the training of singing birds; of the management of domestic birds and the methods of Hatching them by artificial heat. Lastly, we shall enquire concerning those Birds that may be made serviceable to mankind by being reclaimed from their wild state and domesticated.

The third Class in the Linnæan system of Zoology is called

Amphibia

This Class contains a variety of Animals very different in form and in manner of Life. A great part of them inhabit the ocean as the Sharks, Piajas, Sturgeon, etc. which formerly used to be enumerated with the fishes. Some of them on the other hand live on the Land as many of the serpents; and others that are more strictly Amphibious live sometimes on land and sometimes on sea, such are the Lizards, Frogs, Tortoise, etc.

These tribes, tho very different in their aspect and situation yet agree so well in certain characters as made it necessary to include them in one Class.

They have all a single heart with one Auricle and one Ventricle; their Blood is cold and by this they differ from birds, and they breathe thro lungs by which they differ from fishes. Nature seems therefore to have pointed out for them an intermediate station and as they do not class with either the mammalia or fishes to require their constituting a Class by themselves. Unless the Insects they form the most numerous Class of any in the Animal Kingdom, and indeed their livid colours, their cold, naked skin; their reptile motions, their poisonous qualities and the suspicious appearance of such of them as are innocent render them the most detest[a]ble set of Animals to mankind who are more inclined to shun than disposed to examine them. For this reason their history is very imperfect and involv'd in Fable and misrepresentation.

We shall first consider their destructive characters; their different ways of Life and singular manners; of their propagation, perspiration and the opinion of their subsisting long in a dormant state.

We shall examine the facts that have been maintained of their being frozen without the extinction of the vital principle.

The surprizing metamorphosis which some undergo. The fascinating power ascrib'd to serpents and the nature of their poisons.

The fourth Class of Animals is that of the

Fishes

whose natural history is commonly known by the name of

Ichthyology

This in the Linnæan system contains all the Fishes strictly so called.

They have all single hearts; breathe thro gills and are called Oviparous.

After giving an account of the best Authors who have [written] on this subject and of the progress made in it by naturalists, We shall give a description of the external parts of Fishes and explain the Technical terms relating to them.

We shall take a view of the different arrangements of Fishes from Aristotle down to Artedi[86] the great reformer in this particular part of Natural History; from whose labours and after whose example Linnæus reformed the whole system of Natural History.

The manner of Respiration in Fishes by Bronchiae or Gills is peculiar to themselves and belongs not to any other of the Aquatic Animals.

We shall likewise attend to their motions and the means by which they are so exactly poised in the element where they move. Of the functions of the fins.

Of the manner of the Generation of Fishes, a subject still in so much obscurity that we say we know little more than that in this they differ from all other animals.

We shall then enquire into their senses and especially of that controverted question, whether Fishes possess the Sense of hearing.

Of their food; Of their Longevity; Of their situations; Of their solitary or Gregarious manner of life.

[86] Peter Artedi, student of Linnaeus, *Ichthyologia* (1738).

Of their migrations and particularly of their constant change of residence at Spawning; Of their domestication; Of the proper construction and management of Fishponds.

We shall then attend to the Fisheries in our own country forming so considerable an Article in natural Economy. Of the Herring Fisheries; Of the Fishery of Cod, Ling, Fresh Salmon; Mack[e]rel [and] pilchards.

The fifth Class is that of the Insects or what is denominated

Entomology[87]

This Class of Animals is commonly vilified by the undi[s]cerning and inconsiderate, but of all others capable of suggesting the most august views of creative wisdom and of evincing the truth of the maxim, *natura nunquam magis quam in minimis tota.*

In this Class many Species are exceedingly useful to mankind, such as *Cochineal; Cantharides* and *Gum lac,* Honey, Wax and Silk are all the productions of Insects. Many again are highly pernicious; some of them poisonous and some formidable to human pride; our cloaths, furniture and provisions are at the mercy of some, and the cultivated productions of the garden are often laid waste by the attacks of others. Every Crop and Grain both in the field and granary is exposed to be hurt, if not lost by some species of Insects.

Our greatest Forest trees are not secure against their invasions, and our domestic Cattle are dangerously infected and sometimes killed by these Animals.

The Man that could obviate the effects produced by one small Fly in our sugar Islands would deserve and obtain a higher praemium than the Man who should discover the Longitude.

The devastation made by Locusts is with reason accounted one of the greatest human calamities.

The various Animals which occasion the calamities will be carefully delineated when we come to the demonstration of the particular species.

The interest of Mankind is more prejudiced by Insects than by all the large Animals taken together. To a superficial observer the utility of Insects to mankind may seem inadequate to the evils they produce.

With respect to their immediate utility it is without doubt really so, but ultimately the advantages we receive from them are far higher than the evils they occasion; the evils are partial but the good universal, this will appear evident from viewing their importance in the economy of nature; it will be there seen that of them all the superior Animals either immediately or ulti-

[87] A very early use of the word in the classroom.

mately depend not merely for subsistence but even existence, and that without them the whole animal world would be deranged, unhinged and incapable of existing in its present situation, and here we will find the truth of the maxim *Natura nihil frustra.*

On the subject of Insects the Ancients have left us almost nothing; in the last Century their history was investigated by many considerable writers, but Linnæus was the first who fixed to this Class its distinguishing Character by a clear and explicit mark vizt. that of their possessing antennae, organs altogether peculiar to themselves and with which we are so little acquainted that it is supposed that these organs are possess'd of some sense unknown to us.

As Insects exceed in number all the other Animals of the Globe it required uncommon talents to arrange and class them in such a manner as that every individual might be easily distinguished and exactly known. Ray,[88] and Swammerdam[89] and Leuenhook[90] in the last century all wrote largely and well on this subject, but it was not till 1735 that they were so arranged into proper classes, orders and genera with universal characters affix'd to them and appeared in a regular system. In this the celebrated Swede (Linnæus) has the greatest merit and the illustration of his classification will be our principal object.

The mechanism of the external parts of Insects must be first examined and the terms of art explained. Their sexes and manner of propagation will be attended to and their metamorphosis a circumstance tho most common yet most wonderful. The peculiar manner of their respiration not by the mouth but by spiracula dispersed on the surface of the body. The expedient used by nature for the preservation both of the individual and species.

Some other enquiries will naturally occur here as the attachment of some Insects to particular plants. This has always been known but not hitherto sufficiently investigated. Many Insects attach themselves only to one Species of plant in which they exist and deposit their eggs. Many indeed feed on several plants but these plants have always a remarkable affinity to one another in some particular qualities which may lead us to enquire, whether or not this may not indicate the medical properties of these plants?

It is probable that those plants especially which are never touched by Insects may indicate an inherent noxious quality like the fruits in desert countries which are left untouched by Birds.

We shall examine the manner of life of those that live in communities such as Bees, Wasps and Ants; the nature of their association and the regula-

[88] John Ray, 1627–1705, *Observations topographical* . . . .

[89] Jan Jacobz Swammerdam (1637–80), *Biblia naturae: Historia insectorum generalis* (1669).

[90] Anton van Leeuwenhoek (1632–1723), pioneer microscopist.

tions of their government. Of the extraordinary effects ascribed to the Tarantula. Of the success of a project made in France for obtaining silk made by Spiders. Of the phosphorescent qualities of Glowworms, and the particular nature of their light which is neither ardent nor Electrical. The Tribes of Lobsters, Crabs etc. which were formerly called crustaceous Fishes, will properly fall in here as in fact they are more nearly allied to Insects by their external qualities and internal structure than to any other Class. The casting of their shells and the reproduction of them when broken by accident will afford matter for curious speculation, and the power they have of voluntarily divesting themselves of their shells is singular.

We shall then take notice of the horror which some Quadrupeds entertain at the sight of certain Insects to all appearance innocent, but which are exceedingly dangerous by depositing their Eggs, a circumstance which we must consider as one of the most surprising exertions of the powerful principle of Instinct.

Showers of Blood, and waters reckoned to be turned into Blood which formerly used to occasion much anxious speculation, is an appearance proceeding from Insects with which we are now well acquainted.

We shall then carefully take notice of the diseases which mankind is exposed to occasioned by Insects, where it will be proper to explain that principle now established that no insect can produce its species within the human body, which many Physicians ignorant of natural history were once inclined to believe.

The sixth and last Class of Animated nature is denominated by the

Vermes

These are certain tribes that stand still lower in the scale of nature and of Life.

Many of them are so different in their frame from the Animals of the former Classes, so simple in their organization and so confined in their vital powers and motions that they have been called imperfect animals.

By a particular survey of their peculiar structure we shall be led to see wherein their alleged imperfection consists and in what they differ from other animals.

Till last Century very few of them were investigated by Philosophers, and never were they reduced to system till the time of the celebrated Linnæus, who divides them into five orders of which the first is called

Vermes Intestina

These are worms inhabiting the Earth, the Waters and the bodies of Animals, perforating every substance in nature, not only wood but even Stone itself. These minute creatures abject as they are, force themselves upon our notice by their pernicious effects.

We shall make some observations on the remarkable tenacity of life in some of these animals. On such as infest the human body and produce diseases and death in every climate. The IInd order of this Class is the

Vermes Molluscula

These are mostly the inhabitants of the Oceans, and formerly very few of them were known, but lately many of them have been discovered by foreign naturalists.

The land and sea snails; the Cuttle fish; the Medusas or sea jellies which serve as food to whales. The Star-fish and the Echinus all belong to the tribe of Mol[l]uscula.

The III Order comprehends the

Vermes Testacea

Including those Animals that are provided with Shelly or Calcareous coverings; most of these also inhabit the ocean; some the fresh water and others are purely terrestrial. The vast variety, the Symmetry and Colouring of Shells must attract the notice of all such as have any taste or Relish for the Beauties of Nature. Cicero informs us that Scipio and Lolius in their walks amused themselves with making collections of Shells on the sea shore and indeed I think it must be a principal amusement in every age of refinement. The study of Shells is called the science of

Conchology

The Animals themselves so far as yet known beyond to one genus or other of the Vermes Molusca but differ in the figure and other circumstances of the shells or covering. The order of Testacea comprehends all the shells of the Ocean of the fresh waters and of the earth.

The most natural method of arranging the Testacea would be founded on the characters of the Animals that inhabit them but this method is altogether impracticable for the opportunities of examining the marine Animals seldom occur. We are therefore obliged to take the characters from the differences observable in the shells themselves. This is the foundation of the Linnean method which tho in this instance is capable of some improvement may be justly considered as a work of the greatest ingenuity and skill.

Here many curious Phenomena will occur; for instance the Tyrian purple of so great note among the Ancients is the production of an Animal of this order, which has been discovered on the British shores.

The Anatifera shell fish and its supposed transmutation into a bird. The various sorts of shell fish which penetrate and subsist on Rocks. The Byssus of the ancients which was a Lichen stuff produced by a particular species of this order.

We shall consider the Generation and History of Pearls and give an account of the Ship worm formerly ranked by Linnæus among the Vermes Intestina, but now properly placed in the order of Testacea.

The IVth order of the Class of Vermes is called the

Vermes Lithophyta

This order comprehends most of the remarkable and petrified bodies called Corals. These tho' of a stony substance yet participate of Animal nature. They were formerly considered as marine plants and about the beginning of this Century the famous Count Marsigle[91] thought he had discovered the flowers of Corals and that he could distinguish [the] number of petals of which these flowers were composed.

But Mr. Peysonell[92] in 1723 found these supposed flowers to be of an Animal nature and by his further discoveries ascertained Corals to be the works of very minute animals.

These Animals called Polypi have given employment to the researches of many ingenious Philosophers. To trace and examine the different species of the several corals and to enquire into the Economy of these minute animals must afford to every person given to philosophical researches not a little pleasure and entertainment.

The Vth and last order of this class is called

Vermes Zoophyta

This is the lowest and last tribe of animated nature. It consists of bodies which participate both of Animal and Vegetable Life and are by some styled plant-animals. The[y] apparently possess the Roots, Stems and branches of plants, and like these too can be propagated otherwise than by Seeds. But tho rooted in a spot they still possess animal sensation and voluntary motion; such are the various Zoophyta which the ocean produces. But there are others loose and detached, which are endowed with Locomotive powers. These however are still but imperfectly discovered and very little understood.

In this place we will have occasion to examine the animal called the Hydra by Linnæus or the famous Polypus whose regeneration after being divided

[91] Count Marsigli believed that the polyps of coral represented flowers.

[92] Charles de Peyssonel (1700–57), native of Marseilles, concluded that corals were animals. See L. C. Miall, *Early Naturalists* (1912), p. 276.

into different parts was first observed by Mr. Trembley[93] of the French Academy about 1740. [This was] one of the greatest discoveries ever made in the Animal kingdom and which dispelled at once the darkness in which the Corals, the Lithophyta, and Zoophyta had till then been involved.

In this part of our course we shall examine into the nature and properties of the Taenia or Tape Worm, and endeavour to investigate the causes of the grievous Maladies it occasions in the human body and in other Animals.

We shall here examine particularly the Extraordinary resemblance this creature has to the productions of the Vegetable kingdom, that while it lives and grows at one extremity it dies and rots at the other.

We shall next take a view of the pennatulae or sea pens, which are the glowworms of the deep. These by their Phosphorescent qualities Illumine the dark bottom of the Ocean like so many Torches.

The history of the most frightful animal in nature (if the accounts we have received of them are true) will then come under our consideration. I mean the *Furia infernalis* of Linnæus. This minute Creature happily for nature and mankind, if it at all exists occurs only in the northern parts of Sweden. It is said to fall down the Air and instantaneously to enter the body of Men and beasts which it kills with most excruciating tortures in less than a quarter of an Hour.

The accounts we have of this animal from the Swedish Philosophers are extremely imperfect and fall under some degree of suspicion.

An Animal we have of the same nature and genus in our own country tho' I have never had the opportunity of viewing it. It is called in Ga[e]lic Fienlen by the inhabitants of the western Islands. Its direful effects I have often seen in the Hebrides in many deplorable instances; every 4th or 5th person in the island of Jura where it is chiefly found [is] affected with crippled and distorted Limbs.

We will then come to the consideration of

Microscopic Animalcula

being so small that they escape the naked Eye. We will examine such as are found in Vinegars, in Vegetable and Animal infusions and in the Smut of grain. Needham,[94] Baker,[95] M[ü]ller,[96] Swammerdam etc., have all done

---

[93] Abraham Trembley (1700–84), *Mémoires pour servir a l'histoire d'un genre de polypes d'eau douce, à bras en forme de cornes* (Leyden, 1744).

[94] J. T. Needham (1713–81) thought that he had produced spontaneous generation.

[95] Henry Baker studied *Volvox* in 1753.

[96] O. F. Müller, naturalist, *Animalcula infusoria . . .* (1786).

much in this investigation, and yet they seem to have made but an opening for the researches of ingenious men. On this subject much remains to be discovered and especially in what relates to the cause of Infectious diseases. We will then examine the degrees of evidence that renders it probable that the Itch, the Pox, the Plague etc. are all the effects of Animalcula of different species.

In this view of the three kingdoms of Nature which I have now exhibited I have been conducted by many respectable authorities and especially by that of Linnæus.

Had I ventured to follow my own opinion I might have been tempted to pursue a different track. But in Elementary Lectures on a new subject it is dangerous to venture far in unbeaten paths.

My own opinion would have led me to prosecute the following method.

I would first have begun with the investigation of the most simple and Inert Earths. I then would have proceeded to more compound Fossils. I would next have examined those that in appearance and structure approach the nearest to organized bodies. I would then have advanced to Vegetables that in some measure resemble unorganized matter and would have concluded with those which approach nearest to the Animal kingdom.[97]

In Zoology I would have begun with the Zoophyta which participate both of Animal and Vegetable Life, and so have proceeded thro' the various tribes of Animated nature up to Man.

By ascending in this way the Scale of being step by step a more natural and more lasting impression of Fossils, plants and Animals would in my opinion have been obtained; but an attempt of this nature cannot be soon or easily executed.

I have now delivered a sort of Syllabus of what is proposed to be done in the following Lectures on the subject of natural History. But several observations remain still to be made and some enquiries occur which require discussion.

Previous to our beginning the particular view of the several branches of our course it will be proper to give a general view of its Literary History.

But this must be confined to an account of the writings and discoveries of the original and most important authors.

In no science whatever is there a greater quantity of useless and uninstructive books than in natural history.

[97] From the simple to the complex is the method today. This section illustrates that Walker considered all forms of life to be linked in a continuous chain and that no sharp line of demarcation existed between animals and plants or between the most primitive organisms and minerals.

Instead of writing from nature the generality of Authors on this subject have only transcribed from others without adding any new fact or any new reasoning themselves on known facts. So that the whole they have left us is a barren and intollerable [sic] bulk of repetitions.

The elements of Natural history may be obtained by such a course as ours and a taste for it acquired in an University. But neither here, in the Closet, nor in the Museum can any one ever expect to become a thorough naturalist. The objects of nature themselves must be sedu[l]ously examined in their native state, the fields and the mountains must be traversed, the woods and the waters must be explored, the ocean must be fathomed and its shores scrutinized by everyone that would become a proficient in Natural knowledge.

In these pursuits there is this advantage to the naturalist that wherever he goes he has a fund of entertainment before him and can find delight in situations where those who have no taste for this Study would be oppress'd with languor and the *tædium vitæ*. How many sources of useful knowledge [lie] hid from the generality of Mankind. Whoever therefore would be a proficient in Natural History must be himself employ'd in collecting the subjects of it. Many people have a disposition to do this who are little acquainted with the science. Many Gentlemen in our own country and others abroad have an inclination to make collections of Natural Objects but are much at a loss to know what things are proper to be collected and as much difficult[y] how to preserve them. This is the case with many people who are settled in different parts of the world.

I shall therefore make it my business to give you a set of instructions for collecting and preserving the productions of nature.

Of all the subjects in the three kingdoms of nature, fossils are the easiest to collect and to preserve; but to render such a collection useful and instructive, their native situation must be mentioned, their indigenous names set down, their remarkable properties taken notice of, and the uses for which they are employed in the Countries and places where they are found. More apparatus and greater care is necessary in the preparation of plants. I shall therefore acquaint you with the construction and management of the *hortus siccus*. I shall inform you of the best method of preserving and transplanting seeds which requires [sic] particular attention, and is a valuable invention for acquiring foreign plants. But as seeds are apt to fail in long Voyages and living plants can sometimes be had where seeds cannot be obtained, I shall inform you of the method of preserving them at sea and of treating them here in their growing state.

I shall then give you rules to be observed on obtaining and preserving

Animals which are still more numerous and complex. The large Quadrupeds and other Animals are attended with more expense and care in their preservation than most peoples are willing to bestow. They are therefore seldom to be found in collections of natural bodies but are for that reason the more to be wished for and the preservation of them will demand particular attention. No Animals require more care and art in preserving than Birds, and they are capable of being rendered exceedingly beautiful. The method of preserving them was first suggested by a Dutch Collection that appear'd in London in the year 1765.

The sight of this collection set many people in England upon preparing and preserving birds in the same manner and in particular Miss Blackburn[98] and Sir Ashton Lever[99] have arrived at such a degree of perfection as can scarce ever be exceeded. Their skill does not so much consist in the mere preservation as in the attitudes and grouping of the birds, which never fail to affect the Spectators with the greatest pleasure and surprise.

We shall then take notice of the method employ'd in the preservation of Reptiles, Fishes etc., in Spirits. Of the best means of obtaining insects, Shells, Corals and other marine bodies.

Natural History is not at present anywhere more defective than in the products of the ocean. Persons residing on the sea shore even in our own country have it very much in their power to make useful collections of this kind. Every storm throws upon land some particular productions of Animal or vegetable bodies, among which varieties are not unfrequently to be met with. After the proper bodies are collected and prepared there is still another care required of the Naturalist and that is to place them in a proper and scientific manner.

Such a collection is called a museum and must subsist wherever an attention to the study of Nature prevails.

Many illustrious Men even among the ancients have been collectors of Natural Curiosities. Caesar whose abilities in Natural History seem to have kept pace with his talents in war and government was the first who made a collection of Fossils. This he did even in his Wars and his collection was afterwards deposited in the temple of Jupiter Capitol[i]nus.

This example was followed in Italy by many illustrious persons while his empire stood and long after his fall a state for this study again revived with

---

[98] Miss Anne Blackburne, botanist, friend and correspondent of Linnæus, collector of objects illustrating natural history.

[99] Sir Ashton Lever, founder of the Leverian Museum, offered the collection for sale to the British Museum in 1738, but it was declined.

the revival of Letters. The Italian Philosophers were long employ'd in making collections in Natural History before the same spirit arose in the north.

There was no Museum in England till the reign of Charles the Ist when Mr. Tradescant[100] a private Gentleman in London made a very celebrated one, a Catalogue of which still exists.

The Royal Society, which was instituted a little after the Restoration of Chas. the IId. gave to Natural History a lustre and a countenance which made it respectable in England.

Among the many great names which appear'd at this time to do honour to Natural History, that of Sir Hans Sloane[101] deserves the first place. His Travels, His residence in the Capital his large fortune and long life gave him opportunities of amassing such a collection as no private person could hope to equal. At his death the nation thought it proper to take measures for making his private collection a public one and so formed the British Museum. They purchased this valuable repository at the price of 20,000 £, and provided a magnificent house (or rather palace) for its reception at £10,000 more, and it is still supported at the rate of I believe, £1000 per Annum.

No other Museum of any consequence has yet arisen in England, but what the first Lord Littleton foretold in parliament (to whom chiefly the nation owes the British Museum) has already come fully to pass. It has already inspired individuals with a love of the science and rendered England highly remarkable for the encouragement and improvement of useful knowledge. As for our own country (I am sorry to say it) it is farther behind in this respect than any other in Europe except those of Portugal and Spain.

In the last Century there appeared here a most extraordinary Man, the most considerable Naturalist that this Country can boast of, I mean Sir Andrew Balfour,[102] Physician to Chas. the IId, the man who first gave rise to the College of Physicians in this place, to the Botanic garden and to the first appearance of a public Infirmary in this City. He was a th[o]rough proficient in Natural History as far as the study had come in his time, and in his travels abroad made a very large collection of Natural Curiosities which at his death were deposited in this University and a Catalogue of it published by the name of Museum Balfourianum. To this the collection of

---

[100] John Tradescant, Musaeum Tradescantianum, London, 1656.

[101] Sir Hans Sloane, 1660–1753; the Sloane collection was the foundation of the British Museum.

[102] Sir Andrew Balfour, Scottish physician and student of natural history.

Sir Robt. Sibbald[103] was afterwards added and both together formed a Museum as respectable as any University in Europe could at that time boast of. It was kept for many years in the Hall up stairs which is the present Library. It was there I saw it and I may well remember it for it was the view of it which first inspired me with a taste for Natural History when a boy at this College.

Soon after by inexcusable neglect this valuable collection perished and all that remains of it is what is contained in these presses, which I have lately with great labour endeavoured to rescue from the Rubbish. Since that time no publick collection has been formed in Scotland and yet without this Natural History cannot flourish.

For some years past proposals have been spoken of for instituting a Museum here. All seem to approve of the proposal, but general approbation will never accomplish it without the spirit and exertions of individuals. Funds indeed for such an undertaking are scarcely to be found. It will perhaps be expected of me that I should be dispos'd to renew and endeavour to accomplish this scheme. It surely has claim to the patronage of all who have a love to this country, and I can say for myself that what is incumbent on me shall not be wanting.

From the account now given of what is intended in the following Lectures it may appear that there lies before us an extensive range of instructive and ornamental knowledge. What time should be allotted to each particular part of this course, or what time each part of it will require cannot be determined till a trial is made.

In Natural History nothing elucidates so much as a demonstration of the several species, nothing tends so much to illustrate as a view of the bodies themselves. In this science more knowledge may be obtained by the eye than can be convey'd by the ear. I shall endeavour therefore always to exhibit specimens of the objects described tho to do this properly would require a public and extensive Museum. I must therefore content myself with having recourse to my own little private collection which tho considerable enough for my opportunities, yet is altogether inadequate for such a purpose.

We shall particularly always have an eye to the natural History of our own country that we may avoid the imputation of being Lynxes abroad and Moles at home. Our first step ought to be a th[o]rough knowledge of the productions of the land we live on. Exotic rarities excite more surprize and perhaps better gratify curiosity, but our own Country will always afford a more instant fund of entertainment and be of more or less advantage to the public.

[103] Sir Robert Sibbald, 1641–1722, first professor of medicine, University of Edinburgh, 1685.

From this survey I hope you will be sensible how much of this science is misrepresented when it is treated as a trivial and unprofitable study. Nothing surely is more adapted to inform the human mind or better fitted for giving us just and sublime ideas of the Creator, or more enlarged views of his perfections and providence.

Taken in a speculative view its object is to discover new and useful productions in Nature, to promote the cultivation of the various arts; to contribute to the entertainment and improvement of individuals; to advance the public good and to turn the truths of nature and science to the public good.

Elementary Lectures ought chiefly to consist of principles and of the facts observations and experiments on which these principles are founded with an examination of their important consequences. For this purpose all tedious reasoning should be avoided; all useless theories neglected and all conjectures omitted except such as may serve as a spur to studious enquiries.

By such laws I shall endeavour to be guided in my subsequent discourses. But from the Nature of the subject I am sensible I stand much in need so I shall hope for your candour and indulgence.

On this new subject I was willing to deliver a general view of all its parts before I began any particular explanation. That Syllabus was the first work I set about as soon as I entertained the view of the following Lectures. I thought it was necessary to draw the *primæ lineæ* of Natural History in the first place, but when I came to consider in detail the particular branches of it I found much to add and much to alter.

In the following Lectures all the topics which I have enumerated will in some degree be illustrated, but not in the same order they have been laid before you. The arrangement will in many places be altered and I hope by the alteration improved, and many subsidiary subjects of discussion will occur which have not been mentioned.

From the survey we have taken the great variety and extent of Natural History may be easily perceiv'd; its parts indeed are so numerous that they won't suffer us to dwell long on any particular one and from the judgement I have formed I thought it better to take a comparative view of the whole than by a minute investigation of one to leave you altogether uninformed of the rest. It is generally understood that only the elements of a science are to be attain'd at an University and the Elements of this will be sufficient to occupy all the time allotted for our course.

I begin then with

METEOROLOGY

OR

The History of the Atmosphere

This is one of the first subjects that must occur to every one that is studious of Nature.

The oldest writer upon it is Aristotle who composed several books on the history of the Elements, in this he was followed by several of the Greek and Roman authors and by all the Schoolmen who commented on his works or chose him for their model. Many judicious and instructive observations, striking and curious facts are to be found in these writings, but their reasonings concerning them are false and delusive, and when they attempt to ascertain the Causes of the Phenomena which they have observed they never fail to run into absurdity and error. This no doubt was a necessary consequence of a material defect in their Philosophy concerning the nature of the Atmosphere and its essential properties.

It was not till the discovery of the Barometer that the History of the Atmosphere became interesting, and many philosophers such as Boyle,[1] Hawksby,[2] and Hales[3] by their experiments on Air, and the openings into

[1] Robert Boyle (1627–91).
[2] Francis Hauksbee, English physicist.
[3] Stephen Hales (1677–1761), English inventor.

nature which the discoveries concerning Electricity have made, have given to natural history fix'd principles and rational Conclusions and reduced into a Scientific form what was formerly a mass of vague matter with erroneous deductions.

The first property of Air which claims our attention is its Gravity or pressure. This is demonstrated by the Barometer an instrument which has afforded many discoveries and is capable of affording many more both in the history of the Air and of the Earth. It has been long known that air is rendered heavier by Cold and that heat has the contrary effect. To measure the density of the air an Instrument was invented at Paris (in consequence of knowing by experiment that the greatest Cold of France increases the density of the Air 1/5 of its weight and the greatest heat diminishes it 1/7th part) which is constructed on an extensive and exact scale though not yet in common use.

Philosophers to the end of the 16th Century, considered Air as essentially light disposed to mount upwards, and absolutely devoid of weight. Several phenomena caused by the weight of the Atmosphere had been observed previous to the period alluded to, for instance, the ascent of water in pumps, the flowing of water thro' the Syphon and the suction and detention of water in a sponge. But they were totally ignorant of the true cause of these Phenomena which they ascribed to what they called a *horror vacui*. This they supposed to be a certain horror which nature entertained at a vacuum, to avoid which even the heaviest bodies spontaneously ascended to fill the empty space.

The famous Galileo was the first who exploded these barbarous notions and put mankind on the true way of investigation.

He maintained that no heavy body, did or could ascend but by the counterpoise of some other body, and he it was who first ascribed the ascent of water in pumps to the pressure of the air.

Torricelli[4] being improved by his instructions and following his Experiments pursued his discovery; and in 1643 formed the first rough model of the Barometer; it was a water Barometer consisting of a wooden tube 40 feet long open at the ends and immersed in water; after extracting the Air and stopping the upper extremity he found the water to rise and remain suspended at the height of about 32 feet. This however was a very unmanageable instrument, requiring a tread-mill to work it.

To avoid this inconveniency he thought of Quick Silver and a glass tube, and from him this has been since stiled the Torricellian experiment. It was however at that time only employed to ascertain the weight of the Air,

---

[4] Evangelista Torricelli (1608–47).

without drawing from it any other Philosophical inferences. Mr. Boyle afterwards made many experiments with this Instrument. By it he found a great difference in the pressure of the Air at diff't times, and in different places. He observed contrary to what had formerly been believ'd that the pressure of the Air was greatest when it was most clear and serene, and that it was least when the atmosphere was filled with fogs and Clouds. He found this Instrument capable of detecting many minute variations in the Atmosphere and of prognosticating in some degree the changes of the weather, and he was the first who employ'd the Toricellian tube as a weather Glass.

By the Barometer we can now form an idea of the height of the Atmosphere. We know that the comparative weight of Quick Silver to water is 14:1, and that of Water to air as 840 to 1, whence it follows that the suspension of Mercury in the Barometer as 30 inches is equal to the weight of water at about 35 feet. Therefore an Inch of a Cylinder of Quick Silver is equal in weight to 14 inches of a Cylinder of Water of the same base, or to 11,760 inches which is 980 feet of an equal Cylinder of Air; hence a Column of Quick Silver of 30 inches high is equal in weight to a column of Air $5\frac{1}{2}$ English miles in height. This is the height of the Atmosphere when the Mercury stands at 30 inches, were it all equally dense. But as it is denser at the surface of the earth than it is above allowance must be made for its gradual rarefaction. The Astronomers teach us that the Suns light is reflected by the Earth's Atmosphere to the distance of about 50 miles.

Mr. Boyle first invented the Condenser, and compress'd Air to 1/15th; Dr. Halley to 1/100th, Professor Leckman[5] to 1/300th and Dr. Hales to 1/1556th of its pristine bulk, and when Air is so condensed it is very near double the density of water.

——— first increased its change of bulk by heat, Mairotte[6] prosecuted these experiments and Mr. Boyle rarefied it to 10,000 times its former bulk. Whence it follows that the Atmosphere must be variable, now higher and now lower, but the medium height according to Astronomical calculations is about 40 or 50 miles; its exact height however we shall never be able to know.

This was the grand discovery made by the Barometer, for till then Philosophers never knew whether the Atmosphere was 5 or 5000 Miles.

We next consider the state of the Atmosphere as we go higher up, where we find that the gravity of the Air diminishes according to the Altitude of the place. It is much heavier at the level of the Sea for instance, than at the

[5] Or Lehman (unidentified).

[6] Edmond Mariotte (1620–84), French physicist.

top of a Mountain; but the degrees of this gradual decrease of gravity are not fix'd with sufficient accuracy.

Mr. Edmund Halley asserted that as you ascend in the Air in an Arithmetical progression, the density of the Air diminishes in a geometrical ratio, and hence Logarithms were applied to the gradual rarity of the Air, and it was thought to be proportional to the square of the distance upwards. If we admit this the extent of the Earth's Atmosphere is immense, as according to this computation at the height of 35 miles the Air is expanded to 1000 times its bulk. Mr. Borgice[7] was the first who took notice of this.

I am of opinion however that the proportions are not so great as people imagine. When you ascend a Mountain indeed the Air rarefies as you go up, but then the Cold increases and therefore the Air must be proportionally condensed. If then we would ascertain the height of the Atmosphere, we must not only calculate its gradual rarefaction but make allowance at the same time for the gradual increase of cold and all calculators who omit the consideration of both these agents must of necessity fall into error.

I come next to show the Cause of the

RISE *and* FALL *of the* BAROMETER

Tho we know in general that the rise and fall of Mercury in the tube of the Barometer is caused by an increase of the weight of the Atmosphere or its decrease yet the cause which produces this alteration in the pressure of the Air is not yet perfectly ascertained.

Dr. Grew[8] was of opinion that the impregnation of the Air was the Cause of this change. In dry weather Salts which impregnated the Air being only suspended increased its weight and consequently raised the Mercury, and that the pressure was lessened by the solution of those salts in wet weather and consequently the Mercury fell.

Dr. Wallis[9] thought that Mists suspended in the Air increased its pressure. Dr. Halley ascribes the changes in the Barometer to the winds wholly. Others have insisted upon an increase or diminution of the Air's elasticity. But Mr. ——— a late writer alledges on the other hand that moist vapours weaken its spring and that consequently the Mercury must fall. Others insist on an undulatory motion in the superior regions of the Atmosphere by which it becomes of a different height and consequently of a different weight.

These are all unsatisfactory Theories. I will not however add any of my

[7] Possibly Pierre Bouguer.

[8] Nehemiah Grew, *Musaeum Regalis Societatis; Anatomy of Plants* (London, 1682).

[9] John Wallis, *Natural History and Antiquities of Northumberland* (London, 1769).

own, but only observe that Theorists often fail by aiming at too great simplicity. For it is my opinion, that the variations in the gravity of the Atmosphere do not arise from one but from several different causes, and that those variations in the Atmosphere affecting the Barometer are connected with the Electrical state of the Air and that all the changes occasioned in the Mercury of this Instrument depend upon the Electricity of the Air.[10]

We come next to consider what is called

The Range *of the* Barometer

This is that space in the Scale between the highest point the Mercury ever rises and the lowest it ever falls in any particular part of the world.

We are by observations assured that the Range varies according to the Latitude of the place. But the proportion of this variation is not yet accurately ascertained and the cause of the Alteration is still unknown.

In the Torrid Zone the variations are very inconsiderable. On the shore of the South Sea at Peru the French Academicians found that the whole range of the Barometer did not amount to above 3 lines. And at Quito immediately under the line[11] among the Cordeliers[12] the range did not much exceed one line. The same small variation seems to take place in most of the West India Islands. In Jamaica 15 degrees from the Equator the barometer did not change 3 Lines in 6 months, there it is always low and at no time subject to any considerable alteration. An acquaintance of mine who had made the same observations there on the Barometer likewise communicated to me another fact which perhaps may depend on the same causes. Having carried with him to Grenada an Electrical Machine he found it was very often impossible to excite the smallest Electrical appearances, and at no time could he ever collect more of the Electrical fluid than was barely sufficient to produce a small snap. The case seems to be the same I am since informed as far as has been tried all over the West Indies.

Such is the Range of the Barometer near the Equator, but it is found to increase gradually as you approach the Tropics. In the latitude of 15 degrees N. the range is observed to be an Inch. At 45° and 46° N. where it is great-

[10] It was popular to ascribe the cause of many things to electricity, but it is surprising to find Walker falling into this trap when he had such a good knowledge of humidity, temperature, and atmospheric pressure. Walker had conferred with Benjamin Franklin when the latter made a trip to Edinburgh. The study of electricity in the eighteenth century held the attention of men in about the same manner as the study of atomic energy dominates thinking in the twentieth.

[11] Equator.

[12] Walker uses this word in several instances when referring to the Andes or to the North American cordilleran system.

est it does not exceed 3 inches. From thence again towards the Pole it is found gradually to diminish. In our own country the range is generally under 3 inches. Dr. Wallis at Oxford observed it to be only about 2 inches, i.e. from 28 to 30 inches. Mr. Derham[13] at Upminster in Essex found the range there to be 2 inches 0.12.

I had occasion to observe the range of the Barometer in one year at Moffat in 1779 and I found that when it was lowest the Mercury stood at 28:0.2 inches and when highest at 29:0.852 Inches. From this and other observations I conclude that the mean altitude there is 28:0.948 inches. The greatest rise of the Mercury in England that I find recorded was that observed in London by Mr. Graham[14] in December 1721 when it stood at 31 inches and 0.850 of an inch. On the contrary the greatest fall seems to have been that at Townly in Lancashire on the 4th of February 1703 when the Mercury stood at 27:0.39 Inches where it had never been since the year 1665.

Between these extremes there appears a difference of 3 inches and a half, but it is not from these rare and extraordinary appearances that the range of the Barometer is to be estimated.

As we go farther north our observations on this subject grow more scanty but from what we have it appears that the range gradually diminishes even to the pole. At 60° N. L. its utmost extent is not above 1 or 2 inches. At 70° and 75° N. it diminishes to one inch, and at 81° N. it appears to dwindle away to 1/4 inch.

Dr. Halley thinks the greatest difference in the range is from 35° to 55° of Latitude. But by what I have observed I am led to think its range is greater in the north of Scotland than in Edinburgh, and that this has made me often wish that an accurate register of the motions of the Barometer was kept in some of the Scotch Isles. There is a difficulty however which occurs, in fixing the range at any particular place, arising from the imperfection of the Instrument itself; for Mr. Holman[15] at Göttingen observed in different Barometers of the same bore, tho very accurately filled, a considerable difference in their motions at the same place. I find in my notes another observation made by Dr. Cunningham,[16] that in the Island of Chosan on the coast of China in the Latitude 35° 35' N. the range of a Barometer placed 18 feet above the Level of the sea in the year 1721 did not exceed one Inch and 0.900

---

[13] William Derham (1657–1735) made barometric studies in London.

[14] Perhaps the Rev. Patrick Graham (1765–1835), minister of Aberfoyle, Scotland.

[15] Samuel Christian Hollmann (1696–1787), professor at the University of Göttingen, was one of the founders of Göttingen Societät der Wissenschaften.

[16] Dr. James Cunningham, surgeon with the East India Company, made observations on weather during a voyage to China in 1700.

of an Inch. From this observation of the state of the Barometer in China, and as we know it in America under the same parallel of Latitude there is a surprising coincidence. For at Charlestown in South Carolina by observations made on the Barometer by Dr. Chalmers[17] there, its range does not exceed 1.7/100th inches.

All these observations tend to show that the Main Range of the Barometer depends ultimately on the latitude of the place. A single year or two or three is quite sufficient to fix the range in any place, for altho it happens in some years that the Barometer rises higher or sinks lower than can be accounted for or than is usual, yet these are extraordinary.

I shall now consider the

PROGNOSTICS

which the Barometer is said to afford of a Change in the Weather.

From the previous indications concerning the weather which may be obtained by it, it has been called the weather glass. For this purpose it has been deemed by many a very uncertain and erroneous guide, nor is this at all surprizing when we consider how few people are qualified to form a proper judgment of the information it is capable of affording. For this purpose a th[o]rough knowledge of the Instrument and of the principles on which it is constructed, are absolutely necessary, and a strict attention to the most minute vicissitudes of the mercury and the smallest alterations in the weather which few have an opportunity of bestowing. Mr. De Luc's[18] Barometer is as perfect as a portable instrument can possibly be contrived. By [this] instrument the Mercury can be measured to so minute a degree as the 1/500 of an inch, but with these qualifications there is little room to doubt that a person might obtain many useful informations concerning the future changes of the weather, and it is certainly a good prognosticator. The indications it gives are not indeed to be considered as infallible, for its motions are sometimes fallacious and capable of disappointing the most discerning observer, as there are many phenomena depending upon causes with which we are not yet acquainted; but its fallacy is far from being general; on the contrary, tho it is not an absolutely certain guide, yet it is an useful Instrument in the hands of the careful observer.

I shall therefore lay before you canons or rules to be observed with respect

---

[17] Dr. Lionel Chalmers, born about 1715 in Scotland, probably at Cambeltown; tradition gives the University of Edinburgh as the place of his education. He made many meterological observations in Charleston. *An Account of the Weather and Diseases of South Carolina* (London, 1776) is reprinted in the 1940 *Year Book* of the City of Charleston, S.C.

[18] Jean A. De Luc (1727–1817), *Relation de différents voyages dans les Alpes de Faucigny* (Maestricht, 1776); *Lettres physiques et morales sur l'histoire de la terre* (Paris, 1779).

to the Barometer in forming Prognostics with regard to the future state of the weather, which seem to be best established and liable to the fewest exceptions. The Barometer then generally rises previous to Rain weather and on the contrary generally falls previous to and during the continuance of Rain.

These are the principal phenomena attending it, but to form proper conclusions from them, regard must be had to a number of other circumstances.

Thus the Mercury always falls during violent winds (which some have supposed owing to the lateral motion of the Air in winds, diminishing its perpendicular pressure, but this does not appear to me at all satisfactory) but when this is the case it rises with uncommon rapidity after the wind is over. Mr. P_____ observed it rise no less than an inch and a half in the space of 6 hours, after a long storm.

The lowest descent of the Mercury in the tube of the Barometer is during high winds, especially when these are accompanied with violent rain; and it is always at its greatest height with a gentle wind at East or N. East and it scarce ever fails to rise preceding an East wind.

During frost the Mercury is always high especially if the weather be cold. If during frost it ever sinks we may pretty certainly expect either thaw or snow. When the weather suddenly changes to fair or foul upon the first rise or fall of the Mercury it is a sign that such weather will not be of long continuance. If during Rain the Mercury rises two days before a change of weather a continuance of fair weather is to be expected; and in the same manner if it falls 2 days in fair weather, a long tract of Rainy weather will probably follow.

When the Mercury is at the lowest, which in this country is about 28 inches, if it rises it indicates fair weather equally as much as if it had stood at 29 inches or upwards and vice versa.

Rainy or Fair weather will always be found to continue longest when the Mercury is stationary whether it be high or low. But the Barometer is seldom or ever absolutely stationary, but is generally varying through an imperceptible degree of space, sometimes not the 500th part of an Inch.

Sometimes the Mercury falls without any apparent cause. This very often happens in consequence of a violent storm at a distance, or in consequence of Earthquakes as was particularly observed to happen at that of Calabria, which visibly affected all the Barometers of the north of Europe.

When there is occasion to ascertain the future state of the weather for a day or few hours, the best rule is an accurate inspection of the surface of the Mercury in the tube and for this purpose the largest tube is the best. If the Mercury is remarkably convex or concave on the surface it shews itself to be upon the rise or fall, and affords strong presumption of immediate fair or rainy weather.

The Barometer however gives little or no notice of thunder showers or those violent falls of rain which happen at particular places.

In constructing our common Barometer the words marked on the scale of these Instruments by the makers as fair, foul, stormy, changeable, are not only useless, but highly prejudicial and tend exceedingly to diminish the usefulness of the Barometer and to discredit it as a rule for judging of the changes of the weather, as these can only be adapted to the level in which the Instrument is made.

There are some not vulgar but even philosophical errors respecting this subject; thus it has been said that winds blowing in opposite directions accumulate the Atmosphere and increase its weight; that when the current is towards the earth the weight of the Atmosphere is increased; and when it blows from the earth on the other hand the pressure is diminished. These however are errors.

I come now to consider the

MENSURATION *of* HEIGHTS *with the* BAROMETER

An experiment that has a great influence on the Natural History not only of the earth but on the Vegetable and Animal Kingdoms.

Monsr. Pariet[19] informs us that the Barometer was applied for measuring heights in France by Mr. P[ascal][20] in the year 1648, which was indeed but 5 years after the experiment of the Torricellian tube was made known in that kingdom; and I find it was applied to the same purpose in this Country by a Mr. Sinclair[21] of this college in 1661, who published in Holland a treatise entitled *ars gravitatis et levitatis.*

The same Experiment was afterwards much pursued by the French Academicians who by order of Lewis XIV drew a meridian line across the Kingdom of France and it has since been prosecuted by many ingenious philosophers.

The principal Question to be determined here is What column of Air corresponds to a given column of Mercury.

This has been generally answered by experiment, but the conclusions of those who have attempted to decide the point vary so much that the question

[19] Unidentified.

[20] Blaise Pascal (1623–62) tested Torricelli's theory of air pressure on the Puy de Dôme in 1648.

[21] George Sinclair, professor of mathematics, *Ars nova et magna gravitatis et levitatis* (Rotterdam, 1669).

remains undetermined, and unless they are applied with more exactness than they have been hitherto they must be given up.

Previous to our entering on this subject it will be proper to give you a comparative account of the Different measures that have been used at different places by different philosophers in ascertaining the present question.

The Parisian foot was found by Sir Isaac Newton to be the English foot as 1068:1000, wherefore in equal numbers the English foot is to the Paris as 90:96 or as 15:16.

The Swedish foot is to the Paris foot as 1000:1096. So that approaches very near to the English foot.

The Leyden foot is to the English foot as 1390:1350. The measure is very frequently used.

The Figurine Foot used in Switzerland is 1 inch and 8/10 larger than the English foot.

In the year 1696 Dr. Halley made the experiment on the top of Snowden in Wales and he fixed it at 90 feet for 1/10 of an inch of Mercury. About the same time Dr. Derham by very accurate experiments made on the monument at London allowed it to be only 82 feet to 1/10th of an inch of Mercury.

Scheuchzer[22] in 1702 by reiterated experiments in the Alps of Switzerland reduced it to 80 figurine feet to 1/10 of an inch of Mercury. But the same philosopher afterwards found it different in one of the Switzerland Alps, for having first measured the height of a perpendicular mountain by a Cord he found it to be 714 Paris feet high, on the top of this mountain he found the Mercury in the Barometer fell 10/12ths of a Parisian Inch. He therefore allowed 71 Paris feet of Air to the Paris 1/10th of an inch of Mercury which makes that height of Mercury equal to 85 feet English measure of a Column of Air.

Dr. Desaguliers[23] in 1725 made many experiments from which he inferred that 1/10 of an inch of mercury was equal to no less than 96 feet.

Here we may observe that these accounts differ no less than from 82 to 96 feet. Such a remarkable disagreement must render calculation on these principles exceedingly precarious and the mensuration of heights by this Instrument to a great degree uncertain.

I shall now give you the result of some experiments which were either made by myself to obtain further satisfaction on this subject or communicated to me by others on whom I can depend.

[22] Johann Scheuchzer (1672–1733), Swiss naturalist.

[23] Joseph T. Desaugliers (1683–1744) proposed the siphon theory of ebbing and flowing springs.

Those made by myself were executed with the help of an exceeding accurate portable Barometer in the construction of those used by Scheuchzer in his experiments on the Alps.

The first experiment I shall mention was made a good many years ago on the Hill called Logan House Hill which is the highest of the Pentlands. At a considerable distance from the Mountain I made choice of a station which was exceedingly convenient by means of a very extensive plain. Here

Table 1st

1.	Piazza of Holyrood house	135
2.	Nether Bow	210
3.	Second story of the Lands at the Cross of Edinburgh	315
4.	High Steeple of Edinburgh from the base taken with a Cord	155
5.	Summit of Calton Hill	350
6.	Castle of Edinburgh on the 24 Pounder battery	510
7.	Summit of Sal[i]sbury Craig	550
8.	Craig Millar	360
9.	Libberton tower	590
10.	Summit of Craig Lockart Hill	540
11.	Summit of Dalmahoy Hill	680
12.	Summit of Arthur's seat	796
13.	The Bucks stone on Braid Hills	690
14.	House of Woodhouselee	720
15.	House of Green Law	585
16.	House of Auchendenny	554
17.	House of Pennicuick	500
18.	Dalkeith	200
19.	House of the Whim (4/10th above Auchendenny)	884
20.	House of New Hall (15/100 above Woodhouselee)	855
21.	High road at the top of the wood braes above Carlip bridge	990
22.	Summit of Spittal Hill	1360

I made a geometrical observation of the Mountain by the Theodolet and agreed in fixing the height at 1115 feet. On the day following I proceeded to the top of the Mountain where the Barometer stood at 1 inch and 2/10ths lower than in the former station. This experiment has the greater chance to be exact because at 5 hours after when I returned to the station where I had taken the geometrical observation the Barometer stood at the same height it had done before. I therefore concluded from this experiment that 1/10 of an inch of Mercury is equal 92 feet of a Column of Air.

This experiment agrees with that of Dr. Halley beyond what I could have expected. As the Mensuration geometrically was liable to inaccuracies a more certain means is where the height can be measured by a line. I had an opportunity of making an experiment in this way; it was by descending

272 feet in one of the mines at Lead Hills and the depth was measured by a line to an inch. From this I inferred that 1/10 of an Inch of Mercury was equal to 86 feet. But indeed at the time this experiment was made a Nonius's[24] divisor had not been applied to the Barometer, for in such minute observations the nicest eye is not capable of judging with sufficient accuracy.

I made however a more satisfactory experiment on the mountain called Hartfell undoubtedly the highest in the south of Scotland, as I determined its exact height by an actual level and I found it above the ground floor of my house at Moffat 2822 feet, which was equal to 2 inches and 56/100ths of Mercury. By a second experiment it gave 2 inches and 30/100 of Mercury and by a third most to be depended on it gave 2 inches and 33/100. From this therefore it appears that 1/10 of an inch of mercury is equal to 96 feet of Air.

We are next to consider the

HISTORY *of the* ATMOSPHERE *with* *regard to* HEAT *and* COLD

Before the invention of the Thermometer no exact estimate could be made of the degrees of heat and cold in the Atmosphere, otherwise than by its effects, or by the feelings of the human body which are well known to be extremely delusive. But now by means of it we have a perfect standard to judge by from the lowest degree of Cold to the highest degree of Heat.

The invention of the Thermometer like many other useful discoveries has crept into the world while its inventor remained unknown. It has been ascribed to the celebrated Galileo, to a great statesman Father Paul. But the general opinion is in favour of the famous Sanctorius.[25]

The Thermometer however long after its invention remained a very imperfect Instrument while it was constructed with Spirit of Wine and other light liquors. Afterwards oil was used for this purpose by Sir Isaac Newton, and it resists a great degree of Cold and Heat, but Thermometers made with this fluid are very imperfect as it is totally unfit for accurate observation from its viscidity and glutinous nature which makes it adhere to the sides of the tube.

Dr. Halley was undoubtedly the first who proposed a Mercurial Thermometer, and yet this hint lay very long neglected. The first Mercurial Ther-

---

[24] Petrus Nonius, Portuguese mathematician, developed a scale that was later modified as the Vernier scale.

[25] Professor at Padua, studied perspiration.

mometer and was unquestionably made by a Dane, one Baron Reimar,[26] one of the noblemen of the Bed Chamber to his Danish Majesty, about the year 1708; whether he had borrowed the hint from Dr. Halley or not I am uncertain.

But this was even still neglected, till it fell into the hands of two celebrated artists Fahrenheit[27] and DeLisle,[28] who constructed them on different scales and dispersed them over Europe about the year 1736. Afterwards Mr. Reaumur[29] formed one of a different construction. Those of Fahrenheit and DeLisle are used in France, Italy, Sweden and Russia; that of Fahrenheit in Germany, Holland and Britain. Those on Fahrenheit's scale made by our Countryman Dr. Wilson of London are in my opinion the most accurate that ever were executed.

Sir Isaac Newton with his usual sagacity was the first who proposed melting snow and boiling water as the 2 fix'd points of gradation, and to these was afterwards added the heat of the human blood.

In this Thermometer therefore there are three fix'd points.

1st. The freezing point, corresponding to 32° of Fahrenheits scale.

2d. The Heat of the Blood in the human body, marked at 98°, and

3d. That of Boiling water at 212°.

The improvements made in the Thermometer and its general use have now gained us an extensive knowledge with regard to the degrees of heat and cold in the Atmosphere and over the whole globe and nothing is more necessary to enable us to form a proper notion of the different climates as our sensations are not at all to be trusted to in determining the degrees of heat and cold, for sensations are caused by sudden changes from any given state of the air.

I shall now proceed to take a view of the different degrees of Heat that subsist in different Countries as they are ascertained by the Thermometer.

WITH RESPECT *to* HEAT

To begin at home. It is only of late years, and by very few people that the variations in this Instrument occasioned by Heat and Cold have been observed in Scotland with any degree of exactness.

The greatest Summer heats in our Country seem to be from 70° to 78° of Fahrenheit's scale. During the great number of years in which I have had

[26] Unidentified.

[27] Daniel Fahrenheit (1686–1736), German physicist.

[28] Jean Baptiste Romé DeLisle, French chemist and mineralogist.

[29] René Réaumur (1683–1757).

occasion to observe the Thermometer, I never saw it in the shade above +79°, but once which was in the summer of 1779. On the 14th of July in that year the Thermometer at Moffat stood in the full shade at +84° at 2 O'Clock P.M. this may be fairly considered as the greatest heat of our Climate, and is a degree indeed which does not occur but once in a great many years.

The summer heats of the Island of Minorca, which were very accurately determined by the late Dr. Cleghorn[30] of Dublin, seldom raise the Thermometer to 87°.

In several of the American states, as for example in Maryland, during the months of July and August the Thermometer frequently rises to +93. In Pennsylvania to 97°. In Virginia during the three summer months from +85° to +95°. At Charles Town in South Carolina it is often found at +98° in the shade, which is the heat of the human Blood. At Charles Town the Thermometer when buried in the sands of the streets was found to rise in 5 minutes to +108°[?] when the heat of the shade is only +88°.

Dr. Chalmers found at Charles town the Thermometer rise[s] from +90° to +101 and then there were people living in Kitch[e]ns where the heat was as high as +115°.

But in Georgia the degree of heat is greater still for in the shade remote from any reflected heat the Thermometer is often found at +102°.

It would seem from these and other such observations that the greatest heat on the globe is not in Islands nor immediately under the Equator, but seem generally to prevail towards the Tropics, that is, from 15° to 21° or 22° of Latitude, for which phenomenon this reason among others may be given that the Countries under the line have the Sun in the Zenith indeed twice a year, but this at a very considerable distance vizt. of six months. Whereas near the Tropics the Sun is twice vertical in a very short space of time, which must certainly give a great additional heat.

It is for this reason as well as others that the heat of Georgia and the Floridas is much greater than in any of the West India Islands, or in any place near the line. The same thing happens in the East Indies. In the Island of Sumatra thro' which the Equator runs the heat is seldom above +86°, but upon the continent more remote from the line as at Calcutta their frequent summer heat is +93° and it sometimes rises to +100°. But further up the Ganges it frequently rises to +100° or +103°.

We come now to make a few remarks on the intensity of the heat in the

---

[30] George Cleghorn (1716–89), army surgeon, *Observations on the Epidemical Diseases in Minorca* (London, 1751).

Sunshine. Dr. Boerhaave found the Thermometer in the sunshine of the summers of Holland as high as +90° of Fahrenheits scale.

I found it in this Country and in the Highlands of Scotland as high as +116°. Dr. Hales in the south of England found it rise to +140°; Musschenbroek[31] in Holland found it raised even to 156°. Yet in Egypt it is so hot that it is a frequent practice with the inhabitants to roast eggs in the Sunshine. Of this we may judge when it is found that to harden the white of an egg +156° are requisite, and hence it must amount to that degree or upwards. Mr. Aminton[32] found it at Montpelier as high as 212°, the point at which water boils, but this is incredible and most undoubtedly there has been some mistake in the experiment.

We come now to consider the

DEGREES *of* COLD *in* DIFFERENT
PARTS *of the* GLOBE

As we have not such degrees of heat in our climate, so neither have we the same degrees of Cold as subsist in many countries or even with places under the same parallel of Latitude.

No remarks on this subject were ever made as far as I can find in Scotland previous to the year 1740. In the remarkable frost of that year I was informed by the late Dr. White[33] of this place that the Thermometer fell to +6° in Edin[burgh].

In 1768 on the 5th of January at 11 O'Clock at night at New Posso in Tweedale, I had access to see it lower than ever I had heard of it in Scotland, that is at +2°; and I found that this degree of Cold was sufficient to freeze an Egg, an instance of Cold which very rarely occurs in our climate.

The Thermometer has of late years several times been observed to fall lower than in any other part of Scotland; it has been found as low as Zero and as I am informed several degrees below it.

In January 1780 it is said to have fallen below Zero considerably, even to —14° and when laid on the surface of Snow, which without doubt greatly increases the cold it descended to 23 below 0, and —33° or the point at which Spirit of Wine freezes.

But the greatest degrees of cold with which we are acquainted are those

---

[31] Pieter van Musschenbroek, one of the inventors of the Leyden jar.

[32] Guillaume Amontons (1663–1705), French physician, *Remarques et experiences physiques sur la construction d'une nouvelle clepsidre, sur les baromètres, thermomètres, hygromètres* (Paris, 1695).

[33] Gilbert White (1720–93), *The Natural History and Antiquities of Selborne* (1788).

related to Professor Gmelin,[34] observed by him during his residence in Siberia. There even in the Southerly parts of the Country the Thermometer in winter was frequently observed to stand at —58°. At the Russian Fort called Argon[35] in 50 degrees of Latitude, the frost is never thawed even in summer, but remains above a yard and a half deep so as to render the digging of wells impracticable.

At the fort of Keeran[36] in December 1738 the Thermometer of Fahrenheit fell to —99°. On the 27th of Novr. the preceding year it had fallen to —107. On the 15th of January following at Fort Keeran, Professor Gmelin found the same Thermometer an hour after midnight 120° below 0.

Of such degrees of Cold the inhabitants of a temperate climate can scarce form any adequate conception, that last mentioned, vizt. —120° is the highest degree of natural cold on record and it was thought even extraordinary by the inhabitants, and many of the wild beasts of the Country were at that time found frozen to death. Mr. Gmelin himself informs us that he could stand the influence of the external Air only for a very few moments while he went to the door of his house in order to note the degree of his Thermometer.

But these surprizing degrees of descent have of late years been found owing not so much to a proportional increase of Cold as to the nature of the Mercurial fluid itself, which is found liable to a great degree of contraction about the act of freezing.

The observations of the late ingenious Mr. Hutchins[37] made at Hudson's Bay, and those of Mr. Cavendish prove that the congealing point of Mercury is about —40°, and yet several cases that have occurred do not correspond with this, for Mr. Hutchins himself once found the Mercury remaining fluid at —79°.

Mr. Braulin[38] of Petersburg during the severe colds of that place produced cold artificially as low as 56° below 0.

Mr. Maupertuis[39] in Lapland had his Spirit of Wine Thermometer frozen while his Mercurial one stood at —33°.

We come now to consider what is called the

---

[34] J. F. Gmelin, *C. von Linné . . . Natursyst. des Mineralreichs . . .* (Nuremberg, 1777–85).

[35] Argrensk, 51° 48′N., 120°E.

[36] Kirensk, 57° 50′N., 108°E.

[37] Thomas Hutchins, captain in the Sixtieth Regiment.

[38] Joseph Adam Braun, *De admirando frigore artificial: Quo mercurius est congelatus dissertatio* (St. Petersburg, 1760).

[39] Pierre L. Maupertuis (1698–1759), French mathematician and astronomer.

RANGE *of the* THERMOMETER

That is its greatest rise and fall observed during one or more years in one Country or in any particular place.

The Range in Siberia is not under 190° that is from 120° below 0 to 70° above it.

In England the Range may be considered about 92°, that is from —4° to +88°.

In Scotland it may be reckoned about 84° that is from Zero to +84°.

In Charlestown in South Carolina it is about 83, that is from +15° to +98.

The Range in Maryland would seem to be about 81°[?], that is from +8° to +93°.

In the Island of Minorca the Range is surprizingly small, not exceeding 32°[?], that is between +48° and +70°.

Wherever the range of the Thermometer is greatest we may expect the most varied climate, and on the contrary we may always expect the most salutary Climate and the greatest degree of Longevity where the range of the Thermometer is least, and consequently the most favourable situation for people of feeble constitution and consumptive habits. For this reason Lisbon, Montpelier, Nice and Italy are the usual places of resort for such valetudinary people, but these are not to be presumed so well adapted for this purpose as Minorca, for in them the range of the Thermometer is more extensive and of course the vicissitudes of seasons and the degrees of heat and cold greater than in that Island.

But tho in Minorca the range of the Thermometer is only 32°, yet it is still less on the Island of Madeira, which is free also from the hurtful Levantine winds to which Minorca and all the shores of the Mediterranean are exposed. These and other reasons point out Madeira as the most salutary place for Pthisical people which Europe affords.

There is a vast variety between the heat of the Sun and Shade in different climates, the difference is least in Summer and Winter and greatest in Spring and Autumn. It is also greater in Cold countries than in those of a more temperate climate.

It is of great importance to compare the Temperature of noon and night. On the Island of Minorca it is seldom more than 3° or 4°. The most salutary climate is where the Midday and Midnight Heat approach nearest. In Minorca this is least, but the case is very different in the north of Europe. In this Country it is about 30° or 40°.

From experiments it is found that the degrees of heat and Cold of different places are *cæteris paribus* as their altitude above the Sea. Hence the warmth and fertility of valleys and the superior Cold and barrenness in all the higher areas of the country. That the Air is warmer at the bottom than at the summit of Mountains has always been a general observation, but it is not long since the true cause of this was ascertained.

The ingenious Dr. Hooke[40] imagined that the Air near the Earth was warmed by the Earth's surface, but the Air on a high mountain being further removed from the warm surface and also more amply moved and changed by winds for colder Air, was not susceptible of such degrees of heat as the lower parts of the Atmosphere.

But it is now found to proceed from a different cause, vizt. from the different degrees of density of the Air in the different regions of the Atmosphere, for being more dense in the valleys it is not only more susceptible of heat but is more retentive of it than the thinner air at the top of a Mountain.

At a certain altitude over the whole globe the freezing degree of cold subsists perpetually.

Mr. Bourguet[41] the French Academician was the first who made observations on this subject and who called it the point of perpetual congelation and on the Andes fixes it at 15,516 English feet above the level of the sea. But as we proceed from the equator it is at a less and less degree of altitude.

In the temperate Zone it subsists at 13,000 feet; at Ten[e]riffe in 28° of latitude at 10,000, and in mountains of the south of France as in Auvergne at 6,740 feet; on Mount Blanc in the Alps at about 9,000 feet.

In England it is the latitude of 52°. Mr. Kirwan found this point at about 5,740 feet. In the north of Scotland it would seem to be about 4,301 feet; this I am inclined to suppose from the snow being perennial.

From this we find that the different degrees of heat have a most remarkable effect on Vegetation. I am persuaded and indeed it must necessarily follow that vegetation must be at an end.

Accordingly the French Academicians on the Andes immediately under the line found that vegetation was extinguished just above the spot where the degree of perpetual congelation took place, that is at about 14,697 feet.

We know however that this extinction of Vegetation does not take place on any of the Mountains of Britain; on Ben Nevis towards the summit hardly anything but Lichens and other Cryptogamious plants are to be found. Hence we may infer that at about 5,000 feet vegetation would be entirely extinguished.

We have already found by the Barometer that the density of the Atmosphere gradually diminished as we ascend in the Atmosphere, and we will

[40] Robert Hooke (1635–1703), English mathematician and natural philosopher.
[41] Louis Bourguet (1678–1742).

now find by the Thermometer that the heat decreases in the same manner as we ascend.

The mode of ascertaining this is by the heat of boiling water, which is always the same when the Air is of an equal degree of gravity; but it alters considerably when the Air in which the water is boiled is denser or more rare.

This single but important fact was first discovered by the ingenious Fahrenheit and it occurred to him by accident. He found the heat of boiling water greater when the Barometer was high, than when it was low, that it takes a greater degree of heat to boil water in dense than in light Air.

Soon after this Monsr. Le Monniere[42] and Monsr. Meran[43] two French Academicians, by experiments found the heat of boiling water greater at the foot than at the top of a mountain, but in what proportion was left by them undetermined.

In order to obtain some light on this subject and to determine what is the height of the Column of Air which corresponds to the fall of the Mercury in the Thermometer one degree in boiling water, I undertook the following experiment a good many years ago.[44]

It was made upon the paps of Jura the highest Mountains in the Hebrides on the 27th of June 1764.

On that day I filled a Barometer on the shore of the sound of Isla at 7 O'Clock in the morning, and being then placed upon the level of the Sea it stood exactly at 29 inches and 7/10ths of an inch. At 10 O'Clock it continued at the same height, when we set out in order to ascend the Mountain which is one continued steep from the sea shore.

After a difficult ascent of 7 hours we gained the summit and again filled the same Barometer which we now found stood at 27 inches and 1/10, that is 2 inches 6/10 lower than it had stood on the sea shore. Fahrenheit's Thermometer was then put into boiling water on the summit of the Mountain, in a Kettle we had prepared for the purpose, and after a great many repeated immersions we found it to stand constantly and regularly at 207°.

(Sir George Schuckburgh[45] who performed the same experiment on Snow-

---

[42] Probably Louis-Guillaume Le Monnier (1717–99), physician and naturalist, *Observations d'histoire naturelle* (1744).

[43] J. J. Mairan, *Dissertations sur les variations du Baromètre* (Bordeaux, 1715).

[44] This experiment, made while Walker was conducting his survey of the Hebrides for the government, illustrates his use of the scientific method and his approach to the solution of scientific problems in general. Again and again he reminds his students that experimentation is the way to scientific discoveries.

[45] Sir George Augustus William Schuckburgh-Evelyn (1751–1804), *Observations Made in Savoy in Order to Ascertain the Height of Mountains by the Barometer* (London, 1777).

den hill on the 15th of August 1778 found it exactly the same as I had found it on Jura.)

After descending the Mountain we reached the shore not before midnight and here again repeated the experiment; the same Barometrical tube was filled and at one O'Clock in the morning stood on the sea at 29 7/10 Inches, precisely the same height it had stood at the preceding morning at 10 O'Clock, and as the Air continued without any perceptible alteration, I have reason to think we had found the height of the Mountain very near with as much exactness as could be expected with the Barometer. The Thermometer was again put into boiling water and after many repeated trials it stood at 213°, whereas at the top of the Mountain it had only risen to 207°. The Thermometer employ'd was one made by Dr. Wilson[46] of Glasgow and I was therefore certain of its accuracy, and the water with which the experiment was made both at the summit of the Mountain and on the shore was taken from a pure perennial spring.

From this experiment then it would appear that a column of Air of the height of the Mountain was equal to 2 inches and 6/10 of Mercury and assuming 94 feet for each tenth inch of Mercury, will make the height of the Mountain 2444 above the level of the sea which is ——— feet less than half a measured mile. The difference of the heat of the boiling water at the top of the Mountain and on the Shore appears to be in Fahrenheits Thermometer equal to 6° and the height of the Mountain divided by this number gives 407 feet for each degree of the Thermometer.

This experiment has been since performed by different people with results in some cases similar, in others considerably different, that have generally made it less.

But of this single experiment this is the result that for every 407 feet you ascend in the Atmosphere the heat of boiling water will diminish one degree at least for 2, 3 or 4000 feet high.

We come next to consider the different degrees of Heat and Cold in the different climates. The most essential article by which they are distinguished and we shall particularly remark the effects of the great degrees of Cold in our high northern Latitudes.

From what we have already observed of the application of the Barometer and Thermometer to the subject we find that the heat of the Climate depends not merely on its distance from the Sun, but likewise on its elevation in the Atmosphere above the level of the Sea.

It is found that between the summit and bottom of a Mountain in Scotland there is as much and often a greater difference in respect of Heat and Cold,

---

[46] Alexander Wilson, College of Glasgow, 1772, had a foundry in that city.

as between two places situated on the same level at the distance of 20° of Latitude.

There are some tracts of Country in Switzerland upwards of 2000 feet above the Sea level where the inhabitants have plentiful harvests and well ripened fruits, whereas in our climate at such a height we could not propose to have any harvest whatever.

In this respect however the most remarkable place is Quito in Peru under the line. Here the Cordeliers form two great chains of Mountains running south and north thro the vast continent of South America.

The western chain is about 40° of Latitude from the south sea, the eastern chain runs parallel to it at the distance of about 8 Leagues; the intermediate Valley which is Quito, tho a valley is yet 9000 feet above the level of the sea, and the inhabitants live higher in the Atmosphere, and breathe thinner air than any people in the globe.

This valley being in an inland country and within 2° or 3° of the line might be supposed a country scarce to be inhabited.

But the case is far otherwise for it enjoys the most temperate and genial of all climates, owing to its high situation and being inclosed by the Cordeliers. There the inhabitants enjoy a perpetual spring and suffer less vicissitudes of seasons than those of any country in the Globe. In consequence of this the vegetation is uninterrupted, and the seasons are only varied by rain and dry weather, the range of the Thermometer the whole year round not exceeding 6°.

The Thermometer of Reaumur is generally stationary there at about 14° or 15°. The fields enjoy a constant verdure and the trees maintain their foliage without intermission. All the grains and fruits of the temperate as well as of the Torrid Zones grow here. The seasons and Crops are so constant that the Inhabitants have no granaries, and make no provision for times of scarcity. But altho the Climate is so mild yet the people of Quito live within sight of the frozen summits of the Cordeliers where a temperature equal to that of Greenland which is about 75°, 76° or 78° North Lat. perpetually prevails. Thus in the Torrid Zone at an elevation of about 18,000 feet is found a variety of climate equal to 75° of Latitude.

With respect to Heat and Cold there is a signal difference observable between Islands and Continents. In every Zone, Islands are always found more temperate than the adjacent Continents. In the Torrid Zone they are not subject to such violent heat, for instance the Island of St. Helena situated in about 16° of Lat. is of a mild and cool temperature when compared to any places on the continents of Africa or America situated under or near the same parallel of Latitude.

In the northern regions islands neither suffer such degrees of heat or cold as the neighbouring Continents. Islands are in general warmer in winter and colder in summer than the adjacent continents, the surrounding ocean defending from heat in the Torrid Zone and from Cold in the temperate and frigid Zones.

As far as I can find Caesar seems to be the first who remarked this Phenomenon, who indeed every where shews himself to have been a most acute and judicious observer of Nature, and the loss of his Ephemerides cannot be compensated which contain'd his observations relative to Meteorology during his wars in different countries. From his short stay in this Island he could not be perfectly acquainted with its climate, but he was perfectly acquainted with that of France, and when speaking of the Climate of Britain he says *"Caelum Gallico temperatius."* Not only the Heat but the Cold too is greater in France and even in the heart of Germany than in Britain. Tho Britain enjoys in general a moderate temperature yet it is liable to cold Paroxysms.

In the summer of the year 1749 on the 10th of June a saucer of water left in the open Air was found frozen. In July 1748 a shower of snow that had fallen on Pentland hills was not thawed till one O'Clock afternoon. Such degrees of Cold in summer are never known in the Continent.

In the Shetland Isles the climate is as much below that of this Island, as it is inferior to the continent. In these Islands their summer cannot indeed ripen the most common fruits, but then in winter they never feel any exceeding degree of Cold, and even whole winters will pass over without the inhabitants ever seeing snow, and they have never any severe frosts or lying snow.

In like manner we may trace the same effects in different parts of the same Island. South Britain possesses a great extent of Land and Inland country. North Britain is more narrow and every where deeply cut by many arms of the Sea, and more exposed to its influence. On these accounts therefore rather than from any difference in Latitude rises that diversity in Climate between North and South Britain. In England the spring is earlier, the difference in this respect between London and Edinburgh being about 16 days. England has also hotter and more steady summers, and therefore produces earlier and better ripened grains and fruits. Thus the south of England every year produces nectarines, apricots, peaches, figs, grapes, chesnuts and hops all exceedingly well ripened, but these are Fruits that with us very rarely ripen at all or are of a very inferior quality.

But from the same cause there proceeds likewise a greater degree of Cold in the winters of England than of Scotland which is intersected with bays and arms of the sea. This appears in every general frost which prevails over the North of Europe. Of this there are many instances. Thus in the year 1740 the Cold was more intense in England than it was in Scotland, and the frost went off at Edinburgh a full fortnight sooner than it did in London.

In the year 1768 we heard of the death of many Sheep in England by the frost which is a thing never heard of in this country.

There are many authentic accounts of the rending of great trees in England by the frost, as in the years 1683, 1708 and likewise in 1740. In these cases the trunk of the Tree or rather the Sap contained in it instantly bursts asunder with a noise equal to the discharge of a piece of ordnance, which is occasioned by the freezing of the Sap which is capable of a very great and almost irresist[i]ble expansion. The degree of Cold necessary to produce this extraordinary effect I never could learn but have reason to think it was not under —6° or —8°.

This also is an incident never known in Scotland which shews plainly that our frosts are never so severe as in England.

Another uncommon effect of Cold in England of which we know little or nothing is the freezing of rain; this is occasioned by a violent frost near the surface while it rains above. An instance of this happened in Somersetshire in 1762 as described in the Philosophical Transactions; and another in Gloucestershire in the winter of the year 1766. In both of these instances all the trees in that country were so loaded with ice that their greatest arms and boughs were broken down; a gale of wind happening to spring up in the night in the last case vizt. in 1766 occasioned such a noise by the breaking of the branches with their enormous Icicles as alarmed the whole inhabitants of the Country.

Such are the differences in the climates of England and Scotland all of which arise from the one being the great and the other the small end of an Island.

The other differences consist in wind and rain of which Scotland has by far the greatest proportion, of which the narrowness of the country, the height and inequality of the ground are probably the causes.

We come now to consider the

HISTORY *of the* ATMOSPHERE
with RESPECT *to* DRYNESS *and* MOISTURE

This subject would well deserve our notice did it afford us any necessary and well ascertained facts, which however unhappily it does not.

There is a great deficiency here in the Natural History of the Atmosphere, and chiefly for want of a proper Instrument a well regulated Hygrometer by which we might measure exactly the degrees of dryness and moisture that prevail in the Air.

About 100 years ago the celebrated Dr. Hooke attempted to make an instrument of this sort by means of the beard of the Wild Oat (the *Arista* of the *Avena fatua*) which indeed is a very singular sort of substance and well calculated for the purpose as the dryness and moisture of the Air have a surprizing effect upon it. On a rounded joint which it has it makes in some cases several complete revolutions, which vary in proportion to the state of the Air with regard to moisture and dryness. But after all attempts that were made with this Instrument it could never be brought to real use.

Mr. Boyle invented a very ingenious contrivance for this purpose; he made use of a piece of Sponge in the scales of a gold balance which by imbibing the moisture of the Air made the scale gradually preponderate, and to make this perceptible an Index was applied to the Scale.

The Italian Academy del Cimento made use of a very different contrivance. They made use of a Metal or Glass vessel filled with pounded snow or ice by which means the moisture of the atmosphere was condensed in drops on the outside of the vessel, and they judged of the degrees of dryness or moisture by the quantity of water received by that condensation.

Others have made experiments with twisted ropes which do remarkably contract and dilate by changes in the Atmosphere, but they are by no means sufficiently sensible for the purposes of an Hygrometer; others have had recourse to twisted rods of Birch and Poplar.

But after all the instrument used by Dr. Hooke is the best that has been contrived. Every body which contracts or dilates by moisture or dryness is capable in some degree of being a Hygrometer such as Oil of Vitriol, Volatile Alkali, etc. but none of these have been brought to any use.

Of late years however many ingenious men have applied to this subject, particularly Mr. De Luc, the two MM Saussures.

Mr. Saussure constructs his Hygrometer of hair; Mr. De Luc his with whalebone.

The great difficulty is in getting fix'd points to this Instrument as we have in the Thermometer, none such however have yet been discovered for the Hygrometer. The two points necessary in this Instrument are the points of extreme humidity and of extreme dryness.

For fixing the point of greatest humidity Mr. Saussure thinks an inverted receiver standing in water is extreme humidity, but here is the loss that this depends entirely on temperature. Mr. De Luc on the other hand is of opinion that the point of extreme humidity is to be fix'd by plunging the Hygrometer and soaking it in water.

For determining the point of extreme dryness, Mr. De Luc thinks that it may be found by inclosing Quicklime in an inverted vessel that the contained air after the action of the Lime would afford it.

Mr. Saussure thinks the Caustic alkali is the more proper body for this purpose and indeed it is better as having a stronger attraction for humidity.

These philosophers are at present engaged in this subject, and if they are successful it will be doing a very essential service to Natural history.

We come now to consider the judgement formed of the degrees of Heat and Cold by our feelings.

In this respect our Feelings are very precarious, and they do but very seldom correspond tolerably with the state of the Thermometer; as both Heat and Cold are more sensibly felt by any sudden transition from the one to the other. This appearance in our feelings has been attempted to be accounted for by the density and laxity induced in the fibres; the dilatation and contraction of the pores of the skin.

Our feelings of Heat and Cold depend not only on the degree of the wind but on the quarter from which it blows. Thus, in a West wind our feelings of heat and cold correspond nearest to the state of the Thermometer. On the other hand with a South wind, we feel rather warmer than the real state of the Thermometer. When the wind blows from the North we feel much colder than in proportion to the degree at which the Mercury stands.

But the greatest discordance between our feelings and the state of the Thermometer is in an East wind, then we never fail to feel a much greater degree of Cold than what the Thermometer points at.

Our feeling with regard to heat and cold is varied by the strength of the wind. Thus in a still air it always feels warmer than when it blows. We must likewise feel it different according to the density of the Air, the dense air being always the coldest.

Thus water feels colder than air, and Mercury than water tho all of them possess an equal degree of temperature.

On this subject some very nice and instructive experiments have been made of late by Colonel Sir Benjamin Thompson upon the power of different mediums in conducting heat.

He has found by experiment that common Atmospheric Air conducts heat when compared with the Torricellian vacuum as 1000 to 605; and that Mercury possesses this power in comparison of water no less than as 1000 to 313. Hence the Mercury will always appear so much colder or hotter than water tho at the same temperature.

But another important point which Colonel Thompson has discovered is that the conducting power of the air is greatly increased by humidity. Thus, where two parcels of Air are precisely the same by the Thermometer, that which contains the greatest humidity will appear to be much colder from its more readily conducting heat, and indeed this principle well accounts for those pernicious effects resulting from damp Air, damp houses and damp beds, and especially for those very pernicious effects arising from Air replete with dew being brought in contact with the human body.

A peculiar harshness in our feelings is always known to accompany an East wind: this principle discovered by Colonel Thompson bids fairer than any other yet known, to account for this peculiar asperity. It is well known that in an East wind the Atmosphere is heavily loaded with moisture and must on that account appear much colder than it really is.

In former times and indeed of late until some very recent discoveries pure air was always reckoned a simple elementary body.

Van Helmont[47] with great sagacity made a sort of half discovery of several permanently elastic fluids which he termed gases.

This opening however was not pursued till very lately, and indeed all the recent discoveries on this subject may be dated from the celebrated experiments of Dr. Black[48] on fix'd air, and he has now received that well deserved eulogy in a foreign country which has been overlooked in his own, and in France is justly stiled "Le Patriarche de Chymie Pneumatic."

His discoveries led other philosophical men to pursue the subject and Atmospheric Air has now been very accurately analyzed, and is found to be very far from a simple substance.

In the terms of the old Chemistry it is generally agreed to consist of phlogisticated air, dephlogisticated air, and fix'd air. The other substances with which it is most frequently impregnated are to be considered as merely accidental. By some fix'd air or the aerial acid is reckoned an essential ingredient in Atmospheric Air, by others as accidental only.

Dr. Keir in some late experiments has rendered it very manifest that the pure and phlogisticated airs are not only merely mix'd but even Chemically combined. These 2 generally subsist in pretty much the same proportion!!! in the Atmosphere. In favour of Dr. Keir's opinion it may be observed that if pure and phlogisticated airs were merely mix'd, being of so very different specific gravities they would naturally separate and the phlogisticated air

[47] Jean Baptiste Van Helmont (1577–1644), Flemish chemist, *Ortus medicinae* . . . (Amsterdam, 1648).

[48] Joseph Black (1728–99), Scottish chemist.

would ascend in the Atmosphere, which affords a very strong presumption in favour of the opinion.

In consequence of the late discoveries with regard to this it was thought an easy matter to determine the degree of purity in the Atmospheric Air of any place, and for this purpose the Eudiometer[49] was invented and with this by means of Nitrous Air it was thought that the quantity of pure air could easily be known by absorbing the phlogisticated air.

It has now however been tried for several years, and has by no means answered the expectations that have been formed of it.

With respect to the

ELASTICITY *of the* AIR

at different heights in the Atmosphere. It is as yet unknown, but has been supposed in proportion to the density of the air; but to this opinion there are several contradictory experiments.

The effects of highly rarefied air on the human body has [sic] been the subject of dispute. By travellers it has been found that the breathing was much quicker, and sometimes that even blood vessels in the Lungs had burst.

But these assertions have been contradicted. Mr. Buguer[50] and his fellow Academicians lived for some time where the barometer stood at 15 French inches and 7 lines, without feeling any sensible effect. And more lately Baron Reibezar[51] on the top of Mount Aetna particularly attended to this, without being sensibly affected.

Of late years however we have had more accurate accounts of this matter from Mr. Saussure who was the second that had reached the top of Mount Blanc. He had along with him several servants and guides to the number of 19, every one of whom on the summit of the Mountain found themselves sensible of a great difficulty and quickness of breathing. Mr. Saussure however does not mention the proportional increases of respirations in a given time. But he observes that after having remained 4 hours on the top of the mountain the pulse of one of his guides was 98 in a minute, that of one of his servants 112 and his own 100; and at Chamouny at the foot of the Mountain their pulses were respectively 43!, 60 and 72.

The cause most commonly assigned for these two phenomena, difficulty of breathing and quickness of pulse, is the expansion of the dense air between the cavity of the Thorax and the pleura, which most affect[s] the diaphragm and the Muscles of Respiration.

[49] An instrument now used for the volumetric measurement and analysis of gases.

[50] Pierre Bouguer (1698–1758), French mathematician.

[51] Unidentified.

It is likewise argued that it must press the Heart and that viscus will of consequence require greater force to expel the blood, and hence the acceleration of the pulse.

This supposition however proceeds from one of a more doubtful nature and to which we cannot give our assent, vizt. that the dense air of the foot of the mountain remains in the body at the summit but I should rather think they would be in æquilibria.

We proceed now to the subject of

EVAPORATION

On this subject we are but little acquainted, few observations have been made and we have had no register with regard to it.

It is a pity we are so little acquainted with the annual evaporation in different countries.

Dr. Hales attempted to ascertain the quantity evaporated from a surface of water and of earth in winter. For this purpose he chose a dry winter day and found an evaporation of 1/21 of an inch in 9 hours, he also made the experiment during frost on the surface of ice and found it 1/8 of an inch in an equal time. From these experiments the conclusion is that in a dry winter day there were about 2/3 evaporated from the [surface of water] while there was only 1/3 from the surface of ice.

It had always been supposed that the quantity evaporated from wet earth was greater than that from water. But Dr. Hales found that during a summer day the quantity of moisture evaporated from a surface of water is to the quantity evaporated from a surface of wet earth as 10 to 3.

Dr. Watson[52] Bishop of Landaff found in a summer day that from the surface of an English Acre of dry parched grass ground no less than 1600 Gallons were evaporated.

As for the annual evaporation in different Countries it is a pity we are so little acquainted with it.

The first who I can find to have noticed this subject was ———. He in the north of Germany found that 28 English inches were evaporated from the surface of water fully exposed to the sun and wind for a whole year, and a great deal depends on exposure, as Dr. Halley who performed likewise an experiment on this subject in a confined area of Gresham College in the city of London, exposed neither to sun nor wind, found there the quantity of an annual evaporation only 8 inches.

There is a curious and interesting fact respecting Evaporation which was

[52] Richard Watson, Bishop of Llandaff.

sometime ago discovered by the late Dr. Cullen,[53] vizt. that a degree of Cold is always generated by evaporation. He found that every fluid while in a state of evaporation generated a degree of Cold. Before this period however many facts had led to this opinion, as the coolness of Islands, of the sea shores and of Fenny countries, and the fact seems to have been known very early to the inhabitants of the East Indies, where in order to lessen the heat of the houses they hang up wet cloths which by the evaporation taking place from them produces this effect. The fact seems first to have been known to Musschenbroek, but was first published by Dr. Cullen in 1756, and afterwards by Sigr. S———[54] in the Acta Taurinensia in 1759.

From all these it appears that when a sensible Thermometer is smeared with water, with milk, with cream etc., a degree of cold is generated and the Thermometer sinks; and on the other hand when smeared with oil of olives, Linseed oil, oil of Cloves, Petroleum etc., a small degree of heat is generated and the Thermometer rises. There is one fluid which is pretty remarkable in this respect, which neither generates heat nor cold and that is the *Oleum Tartari per deliquium*.

This is a fact of great use in explaining many of the Phenomena of Meteorology.

The usual opinion of Philosophers till of late years with respect to the cause of evaporation, was the following:

They maintained that the Air contained in water on the application of heat was expanded and carried off along with its water vesicles. This was the opinion of Mr. Derham and I find is still held by Mr. De Luc. Withstanding this Dr. Halley was the first who started the idea of a Chemical solution taking place in which he was follow'd by Mr. LeRoi[55] and Musschenbroek. And of late years a very ingenious and elaborate treatise was wrote on the subject by Lord Kaimes[56] in support of this opinion; in which he considers Evaporation as a real Solution, in which water is dissolved in Air; that the air is the Solvent and water the Solvend.

In this opinion he was followed by Dr. Hamilton of Dublin, Dr. Franklin, and by Bishop Watson.

Dr. Hales seems likewise to have been of this opinion as he found that he

[53] William Cullen, professor of chemistry, University of Edinburgh, friend and teacher of Walker.

[54] Chevalier Saluce, "Sur la Nature du fluide elastique . . . ," *Memorie della Reale Academia delle Scienze di Torino.* Ser. 1, Vol. 1 (1759).

[55] Unidentified.

[56] Henry Home, Lord Kames (1696–1782), Scottish leader and friend of Walker.

could increase the evaporation of Boiling water very considerably by carrying a current of Air thro' it.

The opinion is certainly very probable, and readily solves many of the Phenomena of evaporation. But there is one objection to this hypothesis which remains to be obviated, but which to me appears unsurmountable, and that is, that water evaporates under an exhausted receiver, and that Mercury rises up in vapour in the tube of a Barometer which is a very perfect vacuum.

Bishop Watson has doubted this fact, but there can be no doubt of it as it has been established again and again by experiment, and we know very well that Mercury in the Barometer evaporates and when placed in the Sun it will be seen condensed and forming small balls which adhere to the sides of the tube.

Evaporation appears to me to be carried on by a Chymical union and attraction, but very different from one between Air and Water. It is highly probable that the attraction between the Electric fluid and the particles of water is the cause of their separation. By this means their specific gravity is diminished. They become lighter and consequently ascend; and further it appears that there is a repulsion among the particles of water themselves.

Nor is this effect of Electricity so surprizing, when we reflect on its strong attraction for all fluids.

Properly speaking there are only 2 conductors of Electricity, Metals and Fluids; we know that the particles of water are strongly attracted by an Electric Tube, and in this way I would presume that the particles of water are attracted by the Electric fluid in the Air. It is likewise known that evaporation is quickest when the Atmosphere is in the most electric state, that it is promoted by a dry air, and diminish'd by a moist atmosphere, and when water is strongly electrified it rises more copiously, which I have frequently proved, by electrifying sheets of wet brown paper.

The first person who seems to have hinted at the one cause of evaporation was Dr. Desaguliers. He plainly hinted in 1741 that Evaporation was owing in some shape or other to Electricity.

An English Gentleman one Mr. Isles[57] pursued this hint and asserted that the Electrical fluid caused the water to ascend by diminishing its specific gravity. That the Electric fluid does cause water to ascend we know by repeated experiments but that it produces this effect by diminishing its specific gravity is not as manifest. He asserts that all vapours are electrified, and the doctrine has since been adopted by Dr. Franklin, and received great confirmation from his experiments. Of late likewise the same opinion has been

[57] Unidentified.

adopted by Mr. Volta,[58] and it has been found by the experiments of MM Lavoisier and De La Place[59] in 4 Experiments that in one the vapour of water was electrified minus in the other 3 plus.

Further I presume that the union of the particles of water with Electric matter must render their specific gravity less. This would be the effect of a combination even of Inflammable Air with water. It is well known likewise that the Electric fluid communicates a repulsion to the particles of water. When a small stream of water is electrified it is immediately divided and dispersed not only into drops but even into a fog, and when a body of water is electrified it evaporates in as brisk a manner as if a great quantity of culinary fire was applied to it; but certainly Culinary fire is insufficient to cause the ascent and suspension of Clouds in the cold regions of the Atmosphere. Hence we may ask whether they cannot be raised as well as suspended by some other cause than culinary or Solar heat. It is highly probable that the Electric fluid is the sole agent of evaporation, and that Culinary fire is only an assistant cause perhaps by giving aid to the electric fluid, nor can mere cold account for its descent in the Atmosphere.

It seems clear from what I have said that Air cannot be the cause of evaporation; it is also clear that heat cannot be the Cause as it goes on in very great degrees of Cold.

We know nothing therefore that can produce it but the power and properties of Electric fire, indeed this solves all the phenomena of evaporation.[60]

We come now to the consideration of

AQUEOUS METEORS

And first of

EXHALATIONS

Sometimes vapours exist in the Atmosphere without producing Clouds but then they always render it obscure; at other times very copious exhalations will subsist without being perceived, but these always obscure the view in a greater or less degree. The most favourable time for a clear prospect is not a clear sunny day, but when they are flying showers with intervals of sunshine.

[58] Alessandro Volta (1745–1827).

[59] Pierre Simon de Laplace (1749–1827).

[60] Walker was struggling to find an explanation for evaporation phenomena, at a time when the concept of relative humidity had not been formulated. Like many others of the day, when confronted with a mysterious natural phenomenon, he tended to fall back upon the equally mysterious "electric fluid" for explanation. Of course this does not excuse the unwarranted conclusion.

In a hot and sunny day when you are on a Mountain and look down, you see the Atmosphere filled with a bluish mist which obscures the prospect. I have frequently observed upon Mountains when the Land was on the one side and the sea on the other, that this haze was visible above the land but not above the sea, hence it is clear that this haze is owing to land vapours.

The distance to which the naked eye reaches depends partly on the situation of the Spectator, partly on that of the object, and further on the state of the Atmosphere.

Hence, it is pretty certain that the eye can command a more extensive view over sea than over land, as land vapours are more unfriendly to vision than the vapours from the sea which are merely watery.

From the Summit of Hartfell, I have seen the Mountains of Braemar in the head of Aberdeenshire on the one hand and those of Crossfell in the borders of Yorkshire on the other, distant not much under 200 miles.

From the top of the paps of Jura I have had a prospect more distant; on the one hand the Isle of Man in which there are mountains above 1500 feet high, and in the opposite direction the mountains in the Isle of Sky which are upwards of 2000 feet in height, the strait line between these cannot be much under 300 miles so that my prospect was about 150 miles on each side. But when I saw the Isle of Man I perceived that this I could not have done unless over sea, as appear'd from my not seeing the high mountains in Scotland tho' much nearer.

Dr. Gaertner[61] informed me that on the top of Mount Buet in the Alps about 9,000[?] feet high he saw on each side distinctly about 60 Leagues. It is likewise well known that ships at sea descry the Peak of Tenneriffe[62] at 50 Leagues distance. But probably the Atmosphere admits in warm climates more extensive prospects than in ours.

Those who have gone to the greatest heights in the Atmosphere have remarked the Azure of the firmament more intense than in the Valleys, thus the MM Saussures found on the Alps that this blue colour was more intense not only during the day but likewise thro' the night than in the country below. The Clouds are of themselves colourless and receive all their colours from refracted light. In summer they are higher coloured than in winter, and in our latitudes they never assume such vivid tints as in the warmer climates.

Clouds are always brighter at Sunrise and at Sunset than at Midday, because when we view the Sun at any of these former periods we view it thro a greater extent of Atmosphere which is loaded with vapour. The Clouds like-

[61] Josephus Gaertner, *De fractibus* ... (Stuttgart, 1788).

[62] Tenerife, largest of the Canary Islands.

wise are always brighter at Sunset than at Sunrise because then we view the Sun thro' the vapours of the day.

In the Tropics the firmament sometimes assumes a bright green colour. This very rarely occurs in this Latitude. I have however seen it once, it was on November the 13th, 1785 about an hour before sunset. The firmament assumed a pretty strong tint of green, but I could perceive that this appearance arose from bright orange coloured and yellow clouds which mixing with the blue of the firmament produced the green colour, and probably this may likewise be the case in the intratropical regions.

We do not know precisely to what height Clouds ascend. But it is not probable they ascend to heights where great rarefaction takes place; it is probable that in our Climate they do not rise higher than 2 miles, but they probably attain a much greater height in warmer climates. That they never rise higher than 18,000 feet as Mr. De Luc has asserted is certainly an egregious mistake.

We may easily imagine that there is a strong [pull] on the particles of Clouds which keeps them together, otherwise they behoved to be dispelled, dissipated and diffused equally over the whole firmament.

There is an observation of great moment with regard to Clouds, tho' but little attended to, which is that they are suspended in a much lower region of the Atmosphere in Northern Countries than towards the equator. Thus a Mountain in our climate which does not exceed the height of 4000 feet has its top invisible by reason of Clouds the whole year over, and if ever visible perhaps not oftener than 20 times during a whole year; whereas in Switzerland the tops of Mountains double that height are often visible, even Mount Blanc is frequently visible to the Summit, and the Pike of Tenniriffe is often seen by Sailors at 30 or 40 Leagues distance.

The Suspension of Clouds is a very curious subject. They gradually ascend in the Atmosphere till they attain a height where the Air is equal to them in Specific gravity, and there they remain suspended. This is exemplified in Mount Vesuvius and Ætna. The thick smoke of the former ascends in a column till it acquires that height where the Air is equal to its specific gravity, it then spreads out and forms a stratum.[63]

But it is otherwise with Mount Ætna, the smoke of which descends till it

[63] Density stratification of the atmosphere and its significance in controlling the height of clouds or erupted gases from volcanoes is clearly described here. In general Walker's knowledge of the atmosphere, based on his studies with the barometer and thermometer gave him fuller knowledge of atmospheric pressure, density, and the origin of various weather conditions than was common for his day. Obsessed by the experiments of Benjamin Franklin and others, however, he assigned electricity an important function in cloud formation.

comes to Air equally dense with itself which is considerably below the summit.

There is evidently a repulsion which mutually takes place between the particles of vapour communicated to them by Electricity; but there seems also to be an attraction between them. It is by these two powers that the particles of the Clouds are preserved æquidistant. If they had not repulsion they would immediately coalesce, and on the other hand without attraction they would disperse into the immensity of space.

It may be reasonably supposed that Clouds are suspended in the Atmosphere by means of the same power by which they are raised, not by common but by Electric fire. If this were not the case how could they be supported in the Atmosphere in high situations where they are exposed to a great degree of Cold, to which they are frequently obnoxious.

It is plain that the repulsion subsisting between their particles . . . resists the condensation by Cold, and it is capable of resisting even a great degree of freezing [so] that humid fog will sometimes remain at a small height in the Air when the Thermometer close to the surface of the Earth stands as low as $+15.°$.

Dr. Franklin supposes that the Clouds which rise from the Sea are more powerfully supported than those arising from the Land, the reason may be that the one is more electrified than the other, or it may also depend on the latter containing a greater number of Terrestrial particles than the former.

There is one appearance which has been remarked by every kind of observers and that is, that the Mountains attract the Clouds, in consequence of which there is always a greater fall of Snow and Rain upon the Mountains than in the Valleys.

When Clouds are over the ocean they are exceedingly fluctuating, while over land they remain more stationary. This fact is so well known to Seamen, that it has given rise to a common practice at Sea, that of observing the Clouds when they wish to discover land. For this purpose they set the compass to a Cloud and observe its bearing. If it preserves the same situation they conclude it to be suspended over land, but if it fluctuates they then suppose it to hang over the Sea, and in these conclusions they are seldom wrong. I have often seen two Islands with high land in them covered with Clouds while the Sea betwixt them remained quite free from Clouds. In short the fact may be seen every day. It has been generally admitted, and has been supposed owing to the circumstance of the Clouds over Mountains being suspended on a shorter column of Air than when they hang over a valley, and may therefore be more easily drawn aside by the currents of Air near the Mountain.

But there is certainly more in it than this. It is found that any light substance neg[a]tively electrified, approaches a mass of matter electrified positively or $+$. This must frequently be the case between Clouds and Mountains, and the Clouds are often found electrified minus, hence when the Cloud comes within a certain distance of the Mountain it must be attracted; and on the other hand if the Clouds are electrified $+$ and the Mountain — they will still be equally attracted.

This is rendered pretty evident by a simple and easy experiment. If a piece of pure amber be rubbed so as to become electric and if then held to the smoke of a taper newly put out it will immediately draw the smoke all round it, and not only attract but arrest it and form it into the appearance of a cloud.

In this way it is not unlikely that the Clouds restore to the Mountains the Electricity which they obtained from the earth, and thus a constant circulation of the Electric matter is kept up.

We come now to consider the nature of

Fogs

These are considered as of 3 different kinds according to their origin,

vizt.
$$\left.\begin{array}{l}\text{Marine} \\ \text{Terrestrial} \\ \text{Aerial}\end{array}\right\} \text{Fogs}$$

The Marine Fogs seem not to have any thing hurtful in them except what arises from the degrees of their Coldness or moisture. It is the same with the Fogs which have their origin in the higher parts of the Atmosphere. These are to be considered as nothing more than the descent of a Cloud and are only noxious and hurtfull from the same causes with the Marine Fogs.

But it is far otherwise with Terrestrial Fogs, which arising from the surface of the earth partake of the qualities of the soil and are very prejudicial to the health of the Human body. They are frequent in all low and fenny countries and tend much to the detriment of the Inhabitants.

The Terrestrial Fogs are sometimes accompanied with a disagreeable odour. This odour must either proceed from the putrefaction of Animal and Vegetable matters, or from sulphureous vapours which must necessarily render them very prejudicial and the only good of ground fogs is that they presage dry weather.

In the year 1782 we had in this Country the most remarkable fog ever known. It was not merely confined to this country but spread in some degree over England and reached even the northern parts of the continent. It came into Scotland with a North West wind, and subsisted during the whole summer while the Volcanic conflagration in Iceland was going on. We are certain it did not rise immediately from the earth altho' it was merely of a terrestrial nature, and always dry and so thick that the Sun's beams seldom penetrated it. It cut off the usual communication between the Sun and the Earth, which was fully manifested in the unfavourable harvest which followed.

There is little reason to doubt that this remarkable haze proceeded to us from Iceland, a confirmation of which was that more Lightning was observed then, than had been seen for many years before, and it is known that the smoke thrown out from Volcanoes is very productive of Lightning.

The East wind in whatever part of the world it blows is always accompanied by a fog, and the Barometer stands always higher than from any other quarter.

The Cause of fogs is far from being obvious. It has generally been supposed that then the water exhalations are so little attenuated that they do not ascend, and also that the Air is then rarefied and becomes lighter, but neither of these are satisfactory nor does the latter accord with the phenomenon of the Barometer standing high in an East wind, and probably they have nothing to do in the matter. They have likewise been ascribed to the want of heat, but very erroneously since we have fogs in the heat of Summer.

It appears rather more probable that fogs depend on Electricity, that they loiter on the surface and cannot raise or elevate themselves for want of a due proportion of the Electric fluid which we have already seen is the main agent in evaporation.

It is well known that during a fog little Electricity can be excited. This has been generally attributed to the fog itself but is more probably owing to the want of Electricity in the Atmosphere (which is the Cause of the fog) as well as to the moisture of the fog.

All fogs except those merely aqueous, phlogisticate the Air and render it noxious. But it has of late been discovered that nature possesses two resources by which the air thus phlogisticated is again rendered pure vizt. by agitation among water, and by the process of vegetation.

There is only another Phenomenon of this kind which we have to mention and that is the Frost Smock of Greenland.

The account given of it, especially by Egede[64] the Dane who was a missionary there for 25 years; is that in the winter time whenever there happens

[64] H. Egede, *A Description of Greenland, Shewing the Natural History, Situation, Boundaries and Face of the Country* ... (London, 1745).

an opening of the Ice there arises from it an exhalation thicker than any fog, and when one approaches it the cold is so excessive that it singes the face and hands, but what is more remarkable when immers'd in it it proves quite agreeable.

From such an imperfect account it is impossible to assign any cause for this phenomenon.

We next proceed to the consideration of

DEW

Which tho' one of the most common appearances in the Atmosphere yet there are many parts of its history which still remain undetermined.

It has been doubted whether the dew rose or fell. The common and prevailing opinion in all ages has been that it falls from the Atmosphere on the Earth.

This opinion however has been controverted by several philosophers who maintain that it arises from the earth into the Atmosphere. A German philosopher Mr. Gerston[65] not only maintains this but has performed several experiments in order to prove it. For this purpose he exposed during the evening several large plates of Glass at different distances from the ground from an inch to a foot and he found that the dew was only formed on the under surface of the Glass, and that the upper surface remained dry.

He made another experiment to the same purpose, he spread a sheet of paper over a thick board on the ground, one half of the paper was laid on the board and the other half projected over it. He found here also that only that part which hung over the board was wetted with the dew while the other that lay upon it was perfectly dry.

These experiments however seem to be of an equivocal nature, and are very far from proving that the dew ascends from the earth and does not descend from the Atmosphere, that the greater part of it descends is pretty evident but that it may sometimes ascend is likewise true.

Musschenbroek and others have supposed that dew was nothing but perspiration from plants but this is not at all probable.

It has been thought that the fall of the dew extended during the whole night and that it was always most copious a little before sunrise. But on the contrary I have always found that most of the dew falls before two hours after Sun setting.

[65] Probably Christianus L. Gersten, author of *Tentamina systematis novi ad mutationes barometri et natura elateris aerei demonstrandas* . . . (Frankfurt, 1732).

The dew appears most on the Leaves of plants, because they are sooner cooled than the earth.

Dew falls only when the Air is very calm, but when there is the least breeze it never falls. In our Latitudes it is most copious when the days are hottest and the nights coldest.

Dew falls neither frequently nor copiously at Sea because of its more equal temperature. It is likewise more copious in the more serene and warm climates, particularly where there is little or no rain as in Egypt, many tracts of Arabia and the Country of Peru.

In the formation of Dew it seems evidently to have a predilection or superior attraction for some substances, and to avoid others.

There seems to be a different habitude of plants with respect to dew, for all those which have pilous and scabrous leaves attract it copiously while those whose surface is glabrous and oily repel it.

It is also attracted in great quantities by Stones, Glass, Wool, Hair, china, wood and paper. On the contrary it is always repelled by the metals, especially those with a polished surface, particularly by Tin, Silver, gold, Lead, and Mercury. If polished steel be exposed to a dewy atmosphere it will continue dry but if the surface be in the smallest degree rough or rusty then it attracts the Dew copiously.

This predilection of Dew for Electrics and its avoiding Conductors would almost insinuate that it is in some way connected with electricity, particularly as it is much attracted by all Sharp points. This is readily seen in the case of plants. The parts of Vegetables which most copiously attract Dew are the small filaments, the setae or hairs on the surface of the Leaves and all their sharp edges. Thus in a dewy morning you may observe all the sharp points in a field of grass having a dewy drop hanging at them, when all the other parts remain dry.

I know of no experiment unless one made to ascertain the annual depth of dew. This was made by Dr. Hales in Essex who found the depth there in one year to amount to 3 inches and .39 decimals.

Dew when collected is remarkable for its septic property, when exposed to the Sun becoming in a few days entirely putrid which shows that it is strongly impregnated with Animal and Vegetable substances. The hurtful effects of dew on the human body are universally known and in almost every country. In this country they are not peculiarly noxious but when most hurtful are known to be so in a peculiar manner to Rheumatic people and many others. But in many parts of the world in the warmer climates dews are peculiarly noxious.

Many Theories have been laid down to account for the baneful effects of dew on the human body, but the principle lately proposed by Colon'l Thompson seems to be most probable vizt. the great power of moist air in conducting heat.

The Abbè Fortisse[66] describes a peculiar sort of dew of a red colour which happens in Dalmatia which falls at a fix'd season and has the peculiar one remarkable effect vizt. of Blasting all the vines.

We proceed next to the consideration of

Hoar Frost

which is nothing else but frozen dew or fog.

Hoar frost first appears in Autumn when there is no Ice formed on the water, because the Earth at that time is so much heated that the freezing degree will not take place upon it.

But in autumn tho' the earth be very warm yet the Atmosphere is often considerably colder. Therefore the humidity evaporated by the heat of the Sun during the day, is condensed in the cold Air and comes to be concreted into hoar frost. For this reason the hoar frost in high and mountainous countries is greater than in the valleys.

In our climate the Hoar Frost seldom continues the whole day, and when it does we have always the most vigorous degrees of Cold.

Hoar Frost as well as dew is powerfully attracted by light edges and sharp points.

When a Thermometer is put upon Snow it stands some degrees lower than when suspended a few feet in the Atmosphere; this is frequently 2° or 3° and I have seen it 6° lower. This was explained formerly by the supposition of Evaporation taking place from the surface of the Snow and thus generating cold.

Professor Wilson of Glasgow has of late made several ingenious experiments on this subject, and in one case he found the difference no less than 12°. It is remarkable too that this difference is greater in Serene Air and least in foggy weather.

In order to ascertain this difference farther Mr. Wilson had a Circular board 2 feet in diameter filled with Snow and nicely balanced with Scales. [It] was always found to subside and become heavier in serene Air while Hoar Frost was formed, while in hazy foggy Air when no congelation happened the scale rose. When the hoar frost formed the Snow was colder, and when none formed it was warmer than the Ambient air. It follows then that

[66] Alberto Fortis (1741–1803), naturalist and prolific writer on geological subjects; author of *Mineralogische Reise durch Calabrien und Apulien aus d. Italie* (Weimar, 1788).

there is more hoar frost in serene than in foggy Air. Mr. Wilson further found that when he covered the scale with pure dry Siliceous Sand there was more hoar frost formed and the Thermometer stood lower, which may be presumed to have a reason; the Sand exposing a greater number of Sharp points, and thus more powerfully attracting the hoar frost.

This production of Cold during the formation of hoar frost is a singular and curious fact and contradictory to many establish'd opinions. We know that rarefaction is the parent of heat, and condensation of Cold; thus Air suddenly attenuated always generates cold which has always been observed in the Air Pump, for after by a few strokes the Air is rendered much thinner there is always a haze which is evidently owing to the cold generated, rendering the Air unable to retain the vapours it was impregnated with which are condens'd by the cold.

We know on the other hand that heat is produced from the passage of vapour to a fluid state and of a fluid to a Solid as takes place in the churning of Butter; and this likewise happens on the freezing of water which may be cooled to 28° if kept quite still, but on congelation it always starts at 32°.

The same thing takes place on a sudden fall of the Barometer, and when it suddenly falls you may expect a great deal of cold.

The cold produced by Hoar Frost seems an exception to these general rules and very probably points at some undiscovered law of nature.

It is argued however in this instance as well as in the case of the other objection to the general power of Cold in condensation (which is the cold produced by precipitating a Solid from a solution), that the appearance may be accounted for by the following supposition. In the first instance the Air having lost the water it contained, which falls in the form of hoar frost, has its capacity for heat increas'd, and consequently absorbs a quantity of the surrounding sensible heat. In the second instance the solution by losing the Solid acquires the same increased capacity of heat and hence the same effect results.

There is an evident and strong attraction between hoar frost and all sharp points in so much that Mr. Wilson found the tips of iron rails covered with it when there was none on the plain surface and sides, in this it is likewise analogous to dew.

We next come to the consideration of

RAIN

Which tho' so common a phenomenon is yet very imperfectly explained. It is well known that all Mountainous countries are the most rainy. Britain in this respect may be considered as an inclined plane from west to east. Along all the west coast which is the highest land the rain falls in the greatest abundance; this is very well ascertained by the Manufacture of Ropes on the west coast and at Leith, which very much depends on fair weather.

At Leith the rope makers can work 32 days in the year more than at Greenock. It is likewise known in the West Indies that those Islands which have mountains as Jamaica have much more rain than those without them.

This is probably to be ascribed to the attraction of the Clouds and Mountains.

The numbers of days of Rain during the year is an observation that ought always to be made for judging of the climate of any place. Tho' very few such observations have ever been made at any particular place. An account was kept at Leyden in the year 1751 and it was found there to be 163 days of Rain. In Holstein in the year 1767 they were found to amount to 184 and in the year 1768 to 173.

I have found in one year at Moffat hail, snow, and rain amount to no less than 197 days.

The largest drops are always in the lowest regions of the Atmosphere and there are never any very considerable drops in the higher. I had an exceedingly good opportunity of observing this on Mountain Ben[67] in the Isle of Mull about 2000 feet in height. When I descended to the bottom I found the largest drops of rain I had ever seen, whereas Gentlemen that remain'd at the top of the Mountain felt nothing but the small drizzling rain usual in spring showers.

Indeed this must happen for as the drops descend it is impossible but they must coalesce, and it is for this reason that the largest drops fall in summer when the clouds are highest, and the drops of consequence having farther to fall.

Some countries are highly remarkable for being almost destitute of rain. There is a tract of Country along the South Sea for near 16° [?] . . . that is from 30° to 19° of South Latitude, about 400 miles in length from the Sea to the Andes, and 30 Leagues in breadth and which being a dead flat is totally destitute of rain; for this reason tho' plentifully stor'd with trees and producing excellent crops yet the country is quite destitute of verdure except on the banks of the rivers. At Lima so little rain falls that the roofs of the houses are made of slender wood and sometimes even of paper sprinkled with ashes to imbibe the dew.

That it never rains in Egypt cannot be admitted. In lower Egypt it does

---

[67] Probably refers to Ben Buie, 2,354 feet. Ben is simply the Gaelic word for mountain.

sometimes rain, and in middle Egypt, but in Upper Egypt that it does not rain is literally true.

There are likewise some parts in Africa especially about the southern tropic entirely destitute of Rain.

The Annual Depth of Rain

is a circumstance generally taken notice of in registers of the weather. Of all the places where this has been observed the Isle of France seems to enjoy the least rain. By a register kept there for 15 years the rain that fell in a year at Paris was at a medium from 14 to 23 inches, that is in one year it had amounted to 23 inches and in another to only 14.

There are some of the low parts of England which are likewise remarkable for small quantities of Rain. At Upminster in Essex in 16 years the quantity of rain had been annually from 15 to 26 inches.

In Northamptonshire it is found to be from 17 to 27 inches. At London they reckon the medium at 29 inches.

In Northumberland at a medium it only amounts to about 21 inches. In Rutlandshire during one year it only amounted to 25 inches.

The dryest part of Scotland is from Berwick to Edinburgh. At Dunse it only amounted to 21 inches, and was found the same at Haiskhill in this neighbourhood by Lord Alemore.[68] At Glasgow it amounts at a medium to about 30 inches. At Manchester the medium quantity is about 31 inches.

But the most rainy tract we have any account of in Britain is at the Seat of Mt. Townly in Lancashire where there has been kept an accurate register for above a century, and the rain was never found. . . .[69]

[LUMINOUS METEORS: LIGHTNING]

The Attraction of Lightning for Metals is very great, and before it was known that any affinity subsisted between Lightning and Electricity it was remarked that when Lightning broke into a house the plates and vessels of metal were always melted; when into a Church the Church plate. It has been known to melt money in a silk purse and the blade of a sword in a scabbard. What is very remarkable on this head is that when Professor Richmond was killed there were other 4 Gentlemen in the room, most of them nearer the extremity of the conductor than he was. Yet he was the only one killed, and the reason was he had in his waistcoat pocket 40 roubles.

[68] Andrew Pringle, Lord Alemore (d. 1776), solicitor general for Scotland.

[69] There is a break in the manuscript at this point.

Wherever danger therefore is apprehended from Lightning we must keep at a distance from all metals.

We have many instances of Metals melting by Lightning but it was an opinion formerly held by some and more lately by Dr. Franklin that this was owing to what they termed a cold fusion. This was not at all clear, but little could be said against it as an hypothesis till a case happened in England and since that many similar ones abroad. In this instance the Lightning ran along the Iron wire of a bell, which in several places was melted and fell upon the boards of the room which it burnt.

It has been known for a considerable time that Electricity has a remarkable effect on Vegetation.

I was witness to some of the first experiments made on this subject which were performed here by a French Gentleman a Mr. M——— while I was a student here. The Electricity gave them a most surprizing assistance in their vegetation. Of two Myrtles one that was Electrified in a week grew 1 1/2 inches while the other that was not electrified had scarcely grown 1/2 inch.

But this effect of the Electricity gradually wears off and comes at last by repeated shocks to be scarcely discernible and entirely to cease.

In consequence of this effect of Electricity it might be presumed that Lightning would have a similar effect. This has never been tried but I am perfectly convinced in my own mind that it is the case since the year 1783. In the summer of that year hardly a day passed without Lightning, and all our crops were very luxuriant, more so than I had seen them before, and than could have been expected from other circumstances of the weather.

There is a remark which I find made by two Roman writers. One from whom we would not expect it, Juvenal takes notice that Lightning promotes the growth of Mushrooms; the same thing is noticed by Pliny as being well known to the Romans.

We may indeed observe that Mushrooms spring up very rapidly after lightning, but we are generally inclined to ascribe it to the rains that fall about that time. The experiment however might be applied to a Mushroom bed of which we have now many in this country.

Another effect of Lightning is on the flowers of plants. All vegetables, trees particularly, are susceptible of Lightning but more especially when in blossom, about that time particularly when the pollen of *farina fæcundans* is shed.

I have frequently had occasion to observe this, but one night especially which was almost entirely occupied by gleaming Lightning in the case of a luxuriant Cherry Tree which grew on the wall of my house, and was then in full blossom. Next morning there was not a flower but was totally black, and the Leaves were likewise hurt, and it was some years before the tree acquir'd its wonted vigour.

There seems to be one Cause for this extremely probable. The Pollen is known to contain within itself a secondary pollen, which secondary pollen it is presumed likewise bursts and throws out a matter of great tenuity and minuteness, and it is well known to be inflammable, and therefore that there should be a strong attraction between it and Lightning is by no means improbable.

The effects of Lightning on the Human body are similar to those it has on vegetables.

At first when Electricity was applied to paralytic cases the most sanguine hopes were entertained of its success because after the 3 or 4 first shocks the effects produced were very surprizing but it soon ceases to have its effect. Dr. Franklin informed me that after the 7th shock he never knew any benefit derived from it.

After all the great discoveries made in the frame of nature in consequence of the advancement of the science of Electricity and tho' it appears clear as day that the matter of Lightning is electric, yet Mr. Lyon[70] a late writer on the subject and one who in other respects is certainly very ingenious, rejects entirely the Electric explosion as the cause of Lightning and recurs again to the old opinion of its being owing to the inflammation of sulphureous and nitrous exhalations.

The next of the Luminous meteors which we come to consider is the

Aurora Borealis

Somehow or other this meteor has been thought of modern date.

The celebrated Dr. Hooke tells us positively that the Aurora Borealis was first made known and first beheld by Dr. C——— when he published his book ——— in the year 1660.

It is true that in Asia and Greece the seats of the ancient Philosophers this phenomenon is scarcely or at all known. But it occurs pretty frequently in the Southern parts of Europe and was known to the Romans. Pliny tells us "———." This is evidently a description of the Aurora Borealis.

The European writers of the middle ages especially take notice of it from the year 394 down even to the days of Kepler.[71]

In the letters of that celebrated Astronomer we find a particular description

[70] John Lyon (1734–1817), English historian, collected fossils and minerals, studied electricity.
[71] Johann Kepler (1571–1630).

of one that happened in the year 1607. And another is described by Gassendi[72] in 1621.

There was a most remarkable one in England on March 17th, 1716, and which is copiously described by Dr. Halley; and many considered this as its first appearance. But there are even notices of it in several of the old English writers, especially of one that happened January 20th, 1560, described by him as burning spears in the firmament. Yet, notwithstanding all this Pointer[73] a late writer on the subject confidently asserts its modern appearance and affirms that it was not known till after the invention of Gun Powder, and he ascribes it to the explosion of Gun powder in the Atmosphere.

The most remarkable case of Aurora Borealis we have on record was seen at the same time at Moscow, Warsaw, Rome, Madrid, Paris and Cadiz, and to be seen at Moscow and Cadiz at the same time it must have been elevated above 20 Leagues in the Atmosphere.

In this instance observations were made at Rome and Paris by which its parallax was fix'd and Mr. Mairan[74] found it to be elevated to 266 Leagues. Still however he maintained that it was within the bounds of the earths atmosphere.

We have many observations delivered on it by the celebrated Bergman. Of one seen at Upsal he had the parallax fix'd and computed its height at 80 Swedish miles and one that happened the 15th of January that same year he found no less than 130 Swedish Miles.

The Aurora Borealis prevails most in northern climates. Thus it is more frequent in France than in Italy, in England than France, in Scotland than in England, in Sweden than in Scotland, in Iceland than in Sweden, and in Greenland than in Iceland.

In the year 1760 Bergman observed the Aurora Borealis took place in Sweden 53 nights, and in the year 1769 during 51 nights. But it is still much more frequent in Iceland, where they are never free of it unless during the months of June, July, and August in which it cannot be perceived as they have then no night, tho without doubt it happens equally then, yet it is not visible on account of the Sun.

The Aurora Borealis is a phenomenon totally unknown within the Tropics and even considerably beyond them, and in these regions Thunder and Lightning are extremely copious. It does not appear within 35° of Latitude.

[72] Pierre Gassendi (1592–1655), physicist, antagonist of Descartes, professor of mathematics at the Royal College of Paris.

[73] John Pointer, *A Rational Account of the Weather* (Oxford, 1723).

[74] Jean Jacques de Mairan (1678–1771), *Traité physique et historique de l'aurore boréale* (Paris, 1733), p. 59.

We shall now describe the Phenomena or property of this meteor which is the only way in which we can be conducted to its cause.

It appears both in the most Cloudy and Serene sky and more frequently in the north. It seldom appears within 25° of the horizon but always higher. When they are very copious it forms a kind of circular appearance above the Zenith, and then called a Corona always of a diluted flame colour. It appears, disappears, and reappears even in the most serene sky.

In form its appearance is fibrous and sometimes radiated. Its fibres or rays generally diverge from the Zenith to the horizon. It is extremely quick and ombratory in its motions. Some parts move across . . . others which appear stationary, which would intimate that they were at very different heights. One other property of this Meteor is that the Stars are distinctly seen through it.

The Aurora australis was first observ'd by a Frenchman in the year 1714. It was afterwards seen by the Spanish Academicians in going round Cape Horn, and since is particularly described by Dr. Forster[75] in latitude 58°. Forster mentions one thing peculiar, that it is of a whitish colour and he thought that the Stars were to a certain degree obscured by it.

To solve this Phenomenon many different causes have been assigned. The oldest opinion is that it arises from the inflammation of Sulphureous vapours in the Atmosphere.

Mr. Mairan has an ingenious astronomical opinion concerning it, i. e. that it is owing to the Suns atmosphere being attracted, and he supposes this may be attracted to the distance of 3,000,000 miles from this body.

The celebrated Dr. Halley and after him Dr. Hamilton of Dublin were of of opinion that its cause was like that of the Tail of a Comet. But it is to be observed that Dr. Halley was the first who compared it to Electric Light.

Mr. Canton[76] supposed it the Electric matter on its passage from positive to negative Clouds. But we know that it often takes place when there [are] no Clouds, and at heights much greater than our atmosphere.

Dr. Franklin supposed it owing to highly electrified Clouds passing to the Equator, but he afterwards doubted its being at all an electric appearance.

Mr. Lyon a late writer thinks it inflammable matter exhaled into the Atmosphere in the equatorial regions and driven by winds and condensed at the Poles, and thus exploding.

But all these accounts are unsatisfactory. It cannot be presumed as owing

---

[75] Johann G. A. Forster, *Observations Made during a Voyage around the World with Cook*.

[76] John Canton, English physicist, produced positive and negative electricity on the same substance.

to Electric matter in the Clouds, because it most frequently happens when the Sky is most serene, and its height is far elevated above that of the Clouds.

The general presumption seems to be that it is an electric appearance subsisting beyond the earths Atmosphere.

There is no Aurora in the Torrid Zone, and there no Electricity can be excited. Till 30° of Lat. Thunder and Lightning is frequent and there no Aurora takes place. Whereas in the Northern Latitudes where we have little Lightning the Aurora Borealis is most frequent.

It is a frequent Meteor in the northern parts of the temperate Zone, where Lightning is very rare.

In the frigid Zone it is still more frequent as in Iceland where Thunder and Lightning are unknown unless during the Volcanic eruptions.

It is likewise a remarkable circumstance in confirmation of the Aurora Borealis being Electric, that it appears that the atmosphere is most electric when it happens.

Mr. Volta has of late found this to be the case by means of his electric condenser.

It is likewise a considerable confirmation of this that the Stars are seen through it, and it is well known that when the electric fire pervades a transparent medium it refracts not light; and farther the electric flashes in the vacuum of an Air pump are similar to the corruscations of the Aurora Borealis, and likewise this takes place in the Torricellian vacuum; thus on the smallest degree of friction applied to the tube of a Barometer you will perceive corruscations exactly similar to those which the Aurora exhibits. Also during a very bright Aurora the Magnetic needle has been frequently observ'd to vary from the N., and it is known that both Lightning and Electricity have precisely the same effects.

It has sometimes been asserted to me, and particularly by Mr. Short who kept the observatory that he was confident he heard a noise attending it. This indeed would be pretty decisive, but can scarcely be admitted as it is too high to reach our ears.

In investigating the Cause of this phenomenon, there occurs an instance of what will frequently happen in the course of these Lectures, for these experiments and observations now related if sufficiently amplified would exceed the bounds of 2 Lectures which however we could by no means allot to them. But I have already afforded enough to the studious enquirer.

Here I may likewise just take notice of this proposition that the whole matter of Lightning, and the Aurora Borealis is derived from the Earth and to the earth again returned.

The next Meteor we come to treat of is called

Draco Volans

Which is a swift blazing meteor, but which very seldom occurs. It is called by some of the old English writers the fire dragon. In the year 1715 one of these was seen over all Britain, and is copiously described in the Philosophical Transactions. But the most remarkable one within memory and perhaps on record happened in the evening of the 18th of August 1783. In that evening there was one that made its appearance, first in the northern parts of Britain, but was not confined to Scotland, but was seen thro' England and the northern parts of France and Germany. Its course was from North to South, and probably it came from the volcanoes of Iceland which were then in a violent conflagration.

We know that during the conflagration of all Volcanoes an immense quantity of Electric matter is formed as is the case in Iceland where during the eruption of the Volcanoes they are seldom a day without Lightning, while at other times they do not know what it is.

The Draco volans therefore seems to be of an Electric nature.

The next of the Luminous Meteors to be treated of is the

Stella Cadens

This Phenomenon is not confined to the northern climates but was seen and is described by Virgil in his Georgics.

There are two remarkable properties of this meteor that tho it shoots from great heights it observes a pretty constant angle of direction, nearly perhaps exactly one of 60°, and it never reaches the earth but is extinguished at some distance from it.

This Meteor has lately been particularly noticed by Mr. Volta and by him supposed owing to inflammable air fired by Electricity.

But an Electric experiment gives some light into this subject. If the Electric spark passes thro' a rare atmosphere (such a one as the half exhausted receiver of an Air pump) which contains a quantity of Moisture; appearances occur exactly similar to those of the Stella Cadens.

There is an old vulgar idea concerning this meteor and one not only held in this country but likewise abroad. There is a very singular anomalous substance in moist ground, a mere gelatina, and this the vulgar suppose to be the substance of the Shit Star.

But unfortunately we are now so well acquainted with this substance that the opinion seems in the highest degree absurd. It was known long ago to

Paracelsus, who supposed it to contain some peculiar virtue and he called it Nostoc. The substance is now well known to botanists to be a real plant and is called Tremella Nostoc by Linnæus; it is indeed one that seems to be upon the confines of organized and inorganized matter for there is not in its appearance any material difference from a mere jelly yet, it is a real plant and propagates itself by seeds.

The next Meteor is the

COMOZANTS

They are well known to Seafaring people. They are luminous appearances occurring only during the night and most commonly at sea, they hover generally on the spindles of the masts, or where there is any Iron work, and are seen chiefly during a storm or previous to one.

When a single one occurs it was called by the Ancients Helena, when two, Castor and Pollux.

By foreign Sailors such as French and Italian it is called St. [Elmo's] fire.

It is unquestionably an Electric appearance, happening frequently at land as well as at Sea; and in several old authors it is mentioned where balls of fire rested on the tops of Spears during night marches and night attacks.

The last of the Luminous Meteors is the

IGNIS FATUUS

Or as it is call'd in this country Will of the Wisp.

I cannot pretend to deliver any specific description of the Meteor for altho it is well known in general yet no one has ever had an opportunity of particularly examining it; for it is remarked to fly from those who pursue it, and to pursue those who fly from it.

It has been thought a gelatinous phosphorescent matter resident in marshy and fenny countries. It occurs more frequently in England than Scotland, and when it does happen here it is always in mossy or marshy ground.

We come now to another part of our subject, vizt.

WINDS

The old definition of Winds given by Seneca is just as good as any other *"Ventus est aera fluens."*

The general Cause of Winds has been supposed a local rarefaction and condensation of the Atmosphere produced by heat and cold.

We know not as yet of any other Cause being ascertained or even supposed, and undoubtedly if we examine the history of the winds in the different climates we will find every thing tending to prove this.

We find in many parts of the world the winds most exactly regulated by the suns heat and by his course, as on the coast of Malabar the wind blows from the East from midnight to midday, and from the West from midday to midnight.

There is a place likewise in Switzerland mentioned by Scheuchzer, shut up by Mountains, where the winds blows constantly from the West from Sunset to Sunrise and from the East from Sunrise to Sunset.

The same is the case with the trade winds and equinoxial, these do most undoubtedly arise from rarefaction produced by heat. And yet there are reasons to suspect that the density and rarity of the Atmosphere are capable of being altered by other causes than heat and cold.

In this climate our most violent winds do not happen during great heat or cold, for we have seldom any violent gales during the coldest part of winter or the warmest of summer.

But they are generally most violent when there are the most moderate degrees of heat and cold, and when the temperature of the atmosphere has long continued equable.

Among many observations tending to confirm this idea I shall mention but one instance. It happened on the first of March 1787. On that day there was at this place an extraordinary and rapid fall of the Mercury in the tube of the Barometer down to 28 3/10 inches, and it very seldom falls so low here. The consequence was that early next morning a violent gale of wind nearly a hurricane took place which lasted for 48 hours.

From this it is plain that the rarity and density of the Atmosphere may be considerably and essentially affected independent of heat and cold, and from this a violent current take place. But it is not easy to say what these Causes are which independent of heat and cold produce these effects.

I may now take notice of the Force, Velocity, direction and Qualities of the Wind.

With regard to its

FORCE

our winds in the temperate and frigid Zones are never very violent either during great heat or cold.

I was once informed by a Gentleman who happened to pass the most part of the year at the North Cape in Lat. 67°, that during summer the weather was calm and serene, and likewise throughout the winter but in the interim betwixt the two the winds were really tremendous.

The force of winds is generally considerably affected by the vicissitudes of day and night.

In our Climate the East wind generally subsides about sunset.

There is an observation occurs in this country, and which is well founded tho' it seems to be little known to writers on meteorology, nor do I find it anywhere recorded, yet is daily observed by all the maritime inhabitants of Scotland, and that is that the wind always increases with the flooding of the tide and diminishes with the ebb. I am confident that the observation is well founded, but the Cause of the phenomenon I do not pretend to assign.

It would appear that the wind is always more violent the greater height we ascend in the Atmosphere.

We have had occasion to have this point establish'd by means of Balloons. The first time Lunardi[77] ascended here it was at the surface quite calm and he accordingly rose to a considerable height at 1300 feet quite perpendicular, but as he ascended the wind seemed to have more influence on him, and when he crossed the Forth at the height of about 5000 feet (as nearly as I could judge) during all his transit at the surface it was perfectly still, but at that height the wind was so great as to carry him 20 miles in less than an hour.

It is well known that a current of air follows every projectile.

Thus when a Stone or Bullet falls into water bubbles of air make their appearance, which is owing to the Air that followed the projectile.

The current of Air which attends a Cannon Ball is the most violent we are acquainted with. We often hear of people being killed by the wind of a cannon ball especially in the case of the Marquis of Turenne who was at least knocked off his horse by it, tho being an old man he was probably killed by the fall.

There is one gentleman to whom we are much indebted for the information he has given on this subject vizt. Dr. Blanc. When Physician to Admiral Rodney's fleet in the action which took place he was very inquisitive with respect to this and he ordered the Surgeons of the different vessels to make observations concerning it and accordingly he received many very satisfactory reports.

It appeared in general for there were many cases that a cannon bullet passing near any person, the place extending from about 16 to 18 inches was in an instant ——— and quickly became livid.

[77] Vincenzo Lunardi, *An Account of Five Aerial Voyages in Scotland.*

The Captain of the Triton by a cannon ball passing near him had his shoulder much affected and it became livid and swelled.

The buttons of a Sailors Trousers were brushed off by the wind of a cannon bullet. A livid swelling took place in the neighbouring parts attended with great pain and violent strangury which cost two or three months to cure.

Colonel M——— by a bullet which passed near his head was struck down and left nearly insensible.

When the Bullet passed near the Abdomen the person was left without sense or motion, his wounds were very difficultly cured. But when it passed near the Stomach in that case the person dropt down dead without any external marks of injury. This would seem to correspond to Mr. John Hurder's idea that the Stomach is more essential to life than any other part whatever. But at the same time it must be observed that any forcible current of Air passing near the Viscera must there induce greater rarefaction and consequently more powerful effects than on the solids of the body.

With regard to the

Velocity

of the Wind. This has never been determined with sufficient accuracy.

The utmost velocity of a ship is about 18 miles an hour but this can be nothing to that of the wind on account of the resisting medium the water.

It would appear that from the top of a Mountain the velocity of the Clouds might be pretty accurately measured by their shadow. But experiments have never yet been made in this way.

A Brisk wind has been supposed to move only 12 miles an hour, and the strongest gale or storm at the rate of only 60 miles.

But we know in the long flight of Lunardi in England that he was carryed 70 miles in less than an hour; so that probably the greatest velocity of the wind is much superior.

A wind at the rate of 60 miles an hour does not go by 11 times as fast as sound, but there is one current of Air with which we are acquainted that is quicker than sound, it is the greatest velocity of Air yet known, and that is the velocity of Air let into the vacuum of an Air pump. This has been calculated as equal to 1300 feet in a second.

As to the

Direction *of the* Winds

This is always more regular on the ocean than near Land. The winds are likewise considerably directed by the lie of the Land; thus in long extended valleys as in the straths of this country from whatever quarter the wind blows it always moves up and down the valley.

The direction is likewise frequently very different even at small distances when the winds are light, thus it is not uncommon to observe 5 or 6 ships all sailing different courses, but this happens only when the winds are light, for a strong gale governs the whole sea.

When we go about 50° from the Equator it always blows from the West.

This constancy of the West wind has been supposed an eddy of the general trade wind, and indeed it is pretty certain that this wind extends much farther in the Southern than in the Northern hemisphere. In the South side of the Line it takes place even within the tropics for in all the Islands of that latitude the surf is always on the West side. Whereas in the Northern hemisphere this does not take place till about 50° latitude.

The prevalence of West and South West winds in our Climate is well known and was first remark'd by Caesar, and it may be said that over all Britain, trade winds always South West or North East are the great rulers and I may add the great disturbers of our climate; and we feel their effects in proportion as we are near the frontiers.

In many parts of the South of England, as in Cornwall they frequently have the South West wind for 4 or 5 months together, and this is likewise the case on the West coast of Scotland. But in France their winds are almost always from the North West.

We shall now take notice of what have been called the Carrying Clouds. Clouds frequently appear going in one direction while the wind blows opposite at the surface. Nay sometimes there may be perceived two different carrying clouds, while the wind at the surface blows a third way. There is however or may be a deception in this; for when two ranges of Clouds move the same way, and even with exactly the same velocity they may appear to move in different directions owing to the different angles which they make at different distances. To judge therefore of this phenomenon the eye should at the same time have a fix'd object, as the moon, a star, a tree, or a building, etc.

We come now to consider the

QUALITIES *of the* WINDS

In the Torrid Zone the winds are warmer when they blow from the land than from the sea. But in the Temperate and Frigid Zones the converse of this is the case. Of this we have a remarkable instance in Sweden in the northern frontiers of which the north wind which blows over a vast continent is by far the coldest. But in the same Latitude on the coast of north America the North West is the coldest, and at the Cape of Good Hope the coldest wind is the South west.

I may here take notice of that peculiar asperity which always accompanies the East wind and the Cause of which is yet very far from being evident.

We know in general that the East wind is heavily loaded with moisture and that all moist air is a much more ready conductor of heat than dry air and consequently in moist air the human body will find a greater apparent cold than is indicated by the Thermometer.

But this superiority in moisture can scarcely be the sole cause of the peculiar harshness of the East wind because even in some West and Southerly winds which are most extremely moist and attended with thick fogs there is never found that chilling which invariably accompanies the East wind.

The East Wind prevails mostly in the Spring and it has generally been supposed in this country that they occur in the spring in consequence of the melting of the snow in Scandinavia. But this cannot be the reason as the asperity of the East wind subsists after these snows are all melted.

It has been alledged likewise that its harshness arises from its blowing over the extensive cold tract of Scandinavia. But we know that at Upsal in Sweden where their greatest summer heat is in the beginning of June which is much greater than happens in Britain even in the dog days and yet even then if an Easterly wind happens to blow it always carrys along with it its distinguishing characteristic.

The mere moisture of the Air therefore, the melting of the Scandinavian Snow or the blowing over the vast continent of Scandinavia certainly do not account for the peculiar aspect of the East wind. But this seems to be owing to some different and some more general cause. Possibly the great difference every where observable in the East wind from the West may arise from the one blowing in the direction of the suns rays and the other in the contrary direction, for by all accounts this asperity of the East wind is felt more or less in every part of the world.

It is remarkable that over the whole northern hemisphere the south wind generally brings rain, and in the southern hemisphere in all the high latitudes the north wind.

It appears then in general that the wind most productive of rain is that which blows from the equator to the pole.

We come now to consider some particular winds and first the

General Trade Wind

The meridian heat of the Sun in his course from East to West must necessarily occasion a great rarefaction, and which he carries along with him to the West. Hence the more dense air must pursue as it were the course of the Sun, and consequently an East wind must necessarily take place every where between the Tropics. It follows the Sun during the day and keeps moving through the night and if during this it has lost any of its force this is again restored with the Sun next day. Thus it becomes perpetual and is called the general Trade wind.

This wind blows in the northern hemisphere a little to the North, and in the Southern hemisphere a little to the South.

It is liable likewise to many variations. In the Northern hemisphere it reaches even to 40° Lat. It is indeed confined generally within the Tropics but it can often be perceived at 40° N. Lat. But it is exceedingly remarkable that in the Southern hemisphere the regular Trade wind is in most cases bounded by 20° Lat.

The Trade wind varies also considerably in summer and winter. Thus when Ships sail from Britain to the East Indies in summer they meet with the trade wind in 40° latitude but when in winter they do not find it till about 30° Lat.

It varies considerably likewise in different years even at the same season. I have often been informed by the commanders of Guinea ships that sometimes they met with it in Lat. 35° while in others at exactly the same season not before 30° Lat.

It is considerably affected by Land, and is no where regular within 40 or 50 miles of continental land and accordingly when Guinea ships leave the coast of Africa they do not find it till about 60 or 70 miles at sea.

It is apt to be affected by the Sea currents. It loses its force and thus gives way to other winds. Thus along the coast of Peru where a strong current runs from North to South and a strong wind accompanies it in the same direction, there the Trade Wind loses its influence.

We come next to the consideration of what is well known to the inhabitants of most of the West India Islands by the name of

Sea *and* Land Breeze

These take place most regularly in most of the intratropical Islands.

The Sea breeze generally begins about 9 O'Clock in the morning. At first it is very faint but increases till noon, continues with the same force till about 3 O'Clock P.M., and then gradually decreases till about 5 O'Clock evening when it ceases entirely.

On the other hand the Land Breeze begins at 6 O'Clock evening, subsists during the whole night, and dies away at 8 O'Clock in the morning.

Without this alternation of the Sea and land breezes many of the West India Islands would be scarcely habitable, and for the little while between them, betwixt 9 and 10 O'Clock A. M. and 5 and 6 P.M. the heat is most sultry and insufferable.

The Cause of the Land and Sea breeze, and of their regular alternation is pretty obvious.

The Sun as it ascends in the horizon spreads his heat equally on land and Sea. But the earth receives it sooner and reflects it more than the Sea, the Air therefore over the land becomes rarefied and lighter. Consequently the Sea Air flows in upon it in all directions.

In the intervals between the two breezes the Air in both places is in Equilibrio. Consequently this produces a perfect stillness.

On the other hand the Earth cools sooner than the water after sunset; in consequence of which the Air over the land rushes forth over the Sea, and thus produces the Land Breeze.

Hence it happens that Ships can come only in the day time into the harbours of such Islands, by the assistance of the Sea Breeze; and depart only during the night by the help of the Land Breeze.

It is observable that the Land and Sea Breezes take place only in considerably large Islands, and never in small ones; and it is never felt above 3 or 4 leagues from the Island.

The Sea and Land Breeze are remarkable in the West Indies, but very little of it is to be found in the Indian Ocean because there the Islands are liable to the Monsoons; the alternate gentle currents of Air which form the Land and Sea Breezes are thus entirely suppress'd or the gentle gale gives way to the Storm.

The Sea and Land breeze however is not confined entirely to Islands, but is likewise observable on many of the intratropical continental shores.

There are likewise some small intratropical Islands where they do not take place as in St. Helena where the wind generally blows off shore, and in Juan Fernandez where it blows off shore without intermission.

It would seem that they requir'd an equality of day and night and hence they are peculiar to the intratropical regions.

We come now to take notice of certain

Periodical Winds

The Etesian Winds of the Ancients. These blow over all Greece, Asia Minor and Egypt. With regard to their nature and origin, they blow for 40 days in the summer season due north with great force; they were supposed by the Ancients to be owing to the melting of the Snow in the Hyperborean regions; like all other winds they fall during night and this was supposed a confirmation of the opinion.

They are cold but altogether healthy; they blow so regularly and with so much steadiness that the Ancients and even some of the Moderns supposed them the Cause of the overflowing of the Nile which is first dammed up by them and when they cease it comes to overflow.

The next periodical wind is the

Scirocco

It blows every year in Italy and Dalmatia, proceeding from the South. It takes place much about Easter. It brings with it great heat, and no rain. It continues about 20 days but regularly ceases at Sun set. It burns up plants and all verdure, and is most injurious to the crops. In the human body it occasions great Lassitude and weakness. But altho it most frequently takes place about Easter yet if in Italy during the summer the West wind ceases they are pretty certain the Scirocco will take place next day or soon after, and then it is far more pernicious than when it happens in the spring, the heat attending it being very frequently upwards of 80°.

Similar to the Scirocco is the

Solano of Spain

This proceeds from the South East. It prevails chiefly over the whole province of Andalusia. When it continues long it destroys the crops and in the human body produces extreme tenuity; it affects the blood and without proper precautions frequently disorders the Head.

The

Monsoons

come next to be considered.

These are local winds prevailing over all the Indian ocean.

The general trade wind is liable to great variations but in general it blows one half of the year in one direction and the other in the opposite direction. These are called by the Sailors the Monsoons from a Malay word signifying a season.

The Cause of these is evidently the Suns declination towards the Tropics and it blows North East when the Sun is on the southern hemisphere, and South East when it is in the Northern hemisphere. Now the former of these the North East wind blows over an immense continent, the other the South East over Islands and a great ocean. This evidently accounts for the one bringing dry weather, and the other an almost unremitting deluge of Rain.

We shall now take notice of the more remarkable of the

And first of the

LOCAL WINDS

Tornadoes of Ethiopia

For a much fuller account of Tornadoes and Winds in general see the Poem lately published, *Botanic Garden* Part 1st, Note xxxiii.

There is a large tract of the Ethiopic ocean which is always calm unless when these Tornadoes occur.

They take place with thunder and Lightning and last for 3 or 4 hours and subside with much rain and a perfect calm. The region where this happens is exactly opposite to the broadest part of the vast continent of Africa from 4° to 18° N. Lat.

The next of these Local winds to be mention'd is what is called by the English Sailors the

Bulls Eye of Ethiopia

Before it takes place they observe a small black point, or brown spot in the Atmosphere. It gradually thickens and enlarges and when it has attained a certain degree an impetuous wind ensues and does not subside till it is as it were drown'd in a great deluge of Rain.

Another of these Local winds is what the English Sailors call the

Harmatan

This proceeds from the African Land especially from Benin.

It lasts commonly for about 3 or 4 hours. It is unaccompanied with Lightning or thunder, but usually ends in rain.

The Harmatan is of a most scorching quality, not only schrivelling up paper and parchment but even pannels of wood.

The last of the Local winds to be mention'd is the

Samiels of Arabia

which is an Arabian word signifying the wind of poison. They prevail in Arabia, Petraea, Persia, Chaldaea but especially in the country about Bagdad.

They are evidently of a pestilential nature, they are confined to very narrow boundaries, exceedingly irregular in their direction and motions, and appear in general to be the effect of whirlwinds. They generally take place in July and August and proceed from the North West. In some years they do not occur at all, but in others they sometimes happen 3, 4, 5, 8, 10 times in a season. They seldom continue but for a few minutes, they are as swift as Lightning, and fly in narrow and dispersed streams, striking with instant death every man or beast that faces them. Providentially however there is a short warning of their approach, which is a thick haze like a cloud of dust rising out of that part of the horizon from which they proceed, and hence people that are out in the fields are constantly on the watch, and on sight of this immediately seek shelter.

When the Samiel does come they have nothing for it but to lay themselves down with their faces to the ground and their heels to the blast, for to face it would be instant death.

The Arabians say, which is not improbable that it always leaves a strong sulphureous smell behind it; and if it happens to follow the tract of a River or Lake or carried out to Sea, it loses in great part its noxious quality. We are likewise informed that if a leg or arm of those who are killed by it is pulled it separates from the body but this in all probability is merely an exag[g]eration of the Arabians.

We come next to consider

Hurricanes

These are either general or Local. In the temperate climates when one occurs it is generally of great extent. Thus a wind deserving the name of hurricane occurring in the temperate climates is generally felt over the whole or a very great part of Europe; but in the warmer climates hurricanes are very local but perhaps it is in consequence of this that their force is greatly increas'd, for in the West Indies in July, Augt. and Sepr. their hurricanes are infinitely superior to anything of the kind we feel in Europe.

In our climate, strong general winds deserving the name of hurricanes take place either in the winter, late in Autumn or early in spring.

From the Catalogue of those hurricanes which have happened in the North of Europe, and which we have on record, it appears that those which happen'd from the 1st of March to the first of September are to those which happened from September to March only as one to eight, and hence the insurance of Shipping might and certainly ought in some degree to be regulated by this circumstance.

There is some ground for thinking that hurricanes fall out in consequence of great eclipses especially of the Moon, and after large Solar eclipses, particularly in the year 1748, as was observed at Aberdour Castle; before the eclipse the day was warm and calm but when it became central, a gale of wind sprung up nearly a hurricane.

In about 50 years only 3 gales of wind deserving the name of hurricane have occurr'd in this country, vizt. on the 13th of January in 1739, on October 6th, 1756 and on December 31st. 1778, and it is remarkable that these 3 all took place upon the conclusion of nearly a total eclipse of the moon.

With regard to the strength of the wind we have no facts on the subject. This has at all times been very uncertain and the comparative strength of the wind at different times and in different places has never been noticed. Several attempts have been made to construct a proper Anemometer, but all these however are as yet unsatisfactory, and it is an instrument still very much wanted in Meteorology especially in diaries of the weather, not only as determining the direction but also the force and strength of the winds at all times.

We come now to treat of

SOUND

Which is in several respects intimately connected with the natural history of the Atmosphere.

Air is a proper Vehicle for sound for no sound is propagated *in vacuo*.

It has been supposed by some to be produced by an undulatory motion in the Air, by others it is ascribed to tremor, the effects of which indeed can be propagated to a very great distance.

It is remarkable however in whatever way it be propagated whether by an undulatory motion or by tremor, that it affects not in the smallest degree the stillness of the Air; thus when a lighted candle is held near a piece of ordnance when discharged, it affects not in the smallest degree the motion of the flame which remains quite still.

It is alledged indeed that Sounds are intense or remiss according to the rarity or density of the Air. The velocity however of Sound is always the same whether the Air be dense or rare. But yet there is reason to think that Sounds are loudest and propagated farthest when the Barometer is low.

Dr. Franklin I remember was of a different opinion and urged an instance in support of it which happened at Philadelphia. A very large Bell was brought over there which stood for some time on the Wharf and when rung the distance to which it was heard was surprizing, about 6 miles distinctly, but when run in a steeple about 90 feet high, the Sound was not near so loud nor was it propagated to such a distance as when rung near the ground. This he imagined was owing to the difference in the density of the Air at the earth and at the height of 90 feet. But another and a more likely supposition is that when rung near the ground the sound was propagated along the surface of the earth, much better than when in the Steeple. Of this we have an instance in the pilots of the Red Sea. In the Islands of the Red Sea the navigation is very dangerous, the ships therefore by mutual consent fire their guns for pilots at Sunrise and Sunset, the inhabitants are on the watch, and in order to hear more distinctly they lay themselves down with their ears to the earth.

Sound is neither so loud nor so far propagated in rain or snow as during fair weather.

It was supposed always by the Corpuscularians that sound was propagated in the same way as the undulations of water; but there is one fact which does not correspond to this, vizt. as sound always goes with equal velocity it was assumed by the Corpuscularians that bodies of different sizes and with different Momenta when thrown in water produced equal undulations, but this is not the case for we know that the undulations are regulated by the force apply'd.

Of all the Sounds with which we are acquainted the report of ordnance reaches to the greatest distance. Even thunder itself is not near so far heard for it is questionable if the loudest Thunder is heard above 6 or 7 miles. But we have several instances of ordnance being heard to a prodigious distance. When the French besieged Genoa, the noise of the cannon was heard 90 Italian miles. At the Siege of Stockholm the report of the ordnance was heard no less than 30 Swedish miles. But the most remarkable account given of this kind was at the Sea Battle of Solley [Solway] where the report of the Cannon was heard at Shrewsbury which is no less than 200 miles distant in a straight line.
The

VELOCITY

of Sound has often been made the subject of Experiment; but the results of these have been attended with very remarkable differences.

I shall mention several of them in order, beginning with those who have determined its Velocity to be the least and ascending to those who have made it the greatest.

Mr. R——— a French Philosopher long ago by experiments determined it at 560 French feet in a second of time.

Sir Isaac Newton in the first edition of his *Principia* alloted to it only 968 English feet in a second, but in the 2nd Edition of his *Principia* he altered his opinion in consequence of the experiments of Mr. Flamsteadt,[78] Dr. Halley

[78] John Flamsteed (1646–1719), astronomer.

and Mr. Derham. The trials of these Gentlemen made it equal to 1142 English feet in a Second, which is equal to an English mile in 4 3/4 seconds.

The Academy del Cimento fix'd it at 1148 feet in a second.

Mr. Le Condamine at 1175.

Mr. Boyle at 1200.

Mac———1281.

Gassendi at 1473.

Mr. M——— at no less than 1474 feet in a second.

Among these various results however the velocity of sound as determined by Mr. Flamsteadt and Dr. Halley appears the most accurate and most to be depended on.

By determining in this way the velocity of Sound, by means of it the distance between places may be pretty exactly determined as the distance betwixt two Islands may be found, or anywhere where water intervenes as the breadth of large rivers, of Lakes, etc; the distance likewise between the besieged and besiegers, and likewise the distance of Thunder may be pretty nearly determin'd.

We come next to take notice of a remarkable property of sound vizt. that it has always an

EQUABLE MOTION

Whether sounds are great or small they constantly move with the same velocity. Neither is this retarded or accelerated by wind, nor by the direction of the sound. For if a cannon is fired off it is equally heard by those behind it as by those that are in the direction of its Muz[z]le, neither is at all altered in pieces of ordnance by a small or large charge of powder.

Mr. Derham formerly affirmed that the velocity of Sound was not affected by heat or cold, day or night, summer or winter. Of late however Hales in his treatise on sound endeavours to demonstrate by Theoretical reasoning that it moves 30 or 40 feet in a second faster in summer than in winter. But this theoretical opinion stands in direct opposition to positive experiments made by the Academy del Cimento.

Sounds are augmented variously especially by dilated tubes as trumpets, hunters horns, or by the speaking Trumpet an instrument which was invented by Sir Samuel Moreland in the Dutch War. Sound is likewise much increased by acoustic tubes or tubes resembling the form of the human ear and hence likewise are the whispering gallereys.

Sound when reflected forms an echo, and in reflected sound as well as light the angle of incidence is equal to the angle of the reflector. All hard bodies reflect sound briskly, but all soft springy bodies as carpets and cloths reflect it very dully. It is generally supposed that sound when reflected is sent back in the form of a Cone, and therefore to hear an echo in perfection the auditor must be placed in the Axis of the Cone, or as it is called, the *centrum conicum*.

The Echo moves equally quick with the sound itself and by this means the distance of the *Centrum conicum* may be exactly ascertained.

Sound reflected from a plain surface is pretty equal to the sound itself, but it is greatly diminished when reflected by a convex, and greatly increased when reflected from a concave surface.

It is found that the most acute human ear cannot discern above 9 or 10 notes of musick in a second. Supposing it therefore 9, it then follows that there must be 1/9 of a second of interval between the Sound and Echo, so that the least distance perceivable in an Echo is 64 feet.

Dr. Plot[79] and others have observed that an Echo which repeats many Syllables, returns more during the night than through the day, and certainly Echos must vary exceedingly with respect to the state of the Air as to density and rarefaction, moisture and dryness.

We come next to take notice of

BLIGHTS

Which are in fact but little known except by their pernicious effects.

Most Vegetables are liable in some degree to be Blighted, but we are best acquainted with it as happening to Fruit Trees.

By Blights trees are frequently killed, either in part or entirely. The Tree or Branch thus affected withers, decays and dies.

Blights are of two kinds either gradual when it happens in some days or weeks, or sudden, when it takes place in the part of a day, but most frequently during the night.

With respect to the first of these, vizt. Gradual Blights, many superstitious causes have been assigned to account for it.

It was Dr. Hale's opinion that this gradual blight was owing to the [diversion] of the sap by strong boughs by which means the other branches were deprived of their proper nourishment, and no doubt this may sometimes be the case, but in most cases it must be accounted for by other Causes and this does not explain why a whole tree should be blighted.

It is a common opinion tho' it is to be ranked among the vulgar errors that

[79] Robert Plot, *The Natural History of Oxfordshire: Being an Essay towards the Natural History of England.*

these blights are the effect of Insects whose ova are brought along with the East wind.

It is certain that Insects often occasion diseases in plants and sometimes even their entire destruction. But the devastation made by Insects ought to be kept entirely separate and distinct from that caused by Blights.

It is my opinion that either the East wind in Spring or any other wind of a drying nature must be commonly the Cause of gradual Blights; not by obstructing the perspiration of the young flower buds or leaves as Mr. Philip Miller[80] supposed, but by increasing the perspiration beyond what the plant is able to support.

I believe the East wind to be scarcely a more effectual agent in the production of a blight than a strong drying wind from any other quarter. The reason of the common opinion is probably this, that the East winds prevail most in the spring when the Trees are in bloom and when the buds and flowers are most susceptible of injury.

In this sort of Blight the best remedies are shelter and water.

It is found by the constant experience of Gardiners that if their trees are sprinkled with water in which tobacco or black pepper have been infused, or if wet litter be burnt to the windward of the trees, a good effect always follows, which they ascribe to the infusion. But in reality it is merely the effect of so much moisture and humidity applied to the Tree as prevents the effects of too violent a perspiration.

I have never known any Blight of this kind happen in rainy weather, and the youngest leaves and blossoms are never hurt when kept moist or wet.

It is remarkable that the whole vegetable production suffer when attack'd with a strong drying wind and hence the blasting of Crops and the consequence of hurricanes when unattended with rain as frequently happens in the West Indies, and which likewise occurs in our Country tho the effects are generally ascribed to Lightning yet it is more probably owing to an immoderate perspiration such as the plant is not able to supply.

With regard to Sudden Blight. It often kills a plant in the space of night or during part of the day.

In this kind of Blight a few leaves or buds, one or two branches and sometimes an entire tree is killed by it. This sudden Blighting is ascribed by Dr. Hales to be the effect of Sulphureous vapours which sometimes prevail in the Atmosphere.

When this Blight happened in the day time it was supposed by Mr. Miller and others to be the effect of Clouds of vapour acting in the Suns rays like

[80] Philip Miller (1691–1771), *The Gardener's and Florist's Dictionary: Or a Complete System of Horticulture* (London, 1724).

lenses, and that when their focus happened to fall upon any bud or blossom of a tree it was in a short time burnt up.

A late writer on the subject Mr. Reimar[81] supposes it owing to an acid vapour in the Atmosphere attracting the phlogiston of the plant; and he likewise supposes that Lightning is fatal both to plants and Animals, by carrying off their vital phlogiston. But these are merely theoretical suppositions which we do not understand and for which there is no foundation.

The sudden Blight happens most frequently during the night and I think the only Cause is the gleaming Lightning which I formerly mentioned. In short, whatever may be said of the power of the East wind or of Insects, I know of no other causes of Blights than Excessive perspiration and Lightning; and the time when trees are the most liable to sudden blights when they are about to discharge their *farina fæcundans* which seems to be strongly attracted by Lightning.

I am sensible however that the effects of bad Soil, of Climate, of frost, of insects of unskillful pruning are often mistaken from the effects of Blight, from which they are very different.

We come next to make some observations on the

SEASONS

All the world over remarkable Seasons sometimes happen and the temperate is equally liable to these with the frigid and torrid Zones.

In this country there have occurred two or three extraordinary Seasons within this 2 or 300 years.

The earliest remarkable season of which I find any accounts was in the year 1541 when there was a continued frost from January 1st to April 10th.

To this storm the winter of 1683 was exceedingly similar, and such like was that of the year 1709.

Similar to those above was the winter of the year 1715 during which in this country there was not a plough put to the ground on account of the frost from November 15th to February the 20th.

And in 1739 the frost set in on December 24th and continued till March 25th.

But in all these cases the cold was less severe and less lasting in Scotland than in England, or France, and still less in Ireland.

It is a common remark and one for which there seems some reason that we have always fertile summers after these stormy winters. It appears likewise that after them a hotter and steady summer generally takes place than

[81] Réaumur, Reimarus?

after mild winters which it is generally remarked are followed by bad summers. Hence it happens that dearth or famine never occurs in this country after such severe winters, but rather the converse of this takes place, and indeed in general there was never a dearth in this country in consequence of a hot summer, but rather from cold and wet summers and rainy autumns. Such were the years from 1695 to 1701, and of late the years 1782 and 1783 seem to [have] been of a similar kind. The summers of these were wet but the winters mild.

It has been supposed by many that there is a Cycle or Circle of seasons, and that they renew their progress again after the expiration of the circle preceding.

The history of the weather in Europe does not all correspond with this opinion.

There is a late Italian writer on the subject Sigr. T——— who endeavours to make a very short Cycle. He is of opinion tho with Little apparent reason, that this Cycle is only 18 years. But there are very few people who have not been witnesses that this is not the case.

In the temperate Zones we have four pretty equal seasons. In the Frigid Zone there are only two seasons; thus in the North of Sweden and Hudsons Bay they have only summer and winter with little or scarce any intermediate weather.

In Lapland the summer is divided from the winter by about two weeks at farthest. Linnæus that year in which he was in Lapland, took notice that in Lat. 69° on the 4th of June the trees began to bud, and in less than a fortnight, they and all other Vegetables were in full flower.

The seasons of the Torrid Zone differ more suddenly than in the temperate, but they are estimated by the inhabitants as 4, two dry seasons each of which are six weeks before and six weeks after the equinoxes; the other two are rainy but remarkably more healthy.

It is natural to suppose that epidemic diseases depend very much on these vicissitudes of Season, and to prove this there are not wanting instances.

In the temperate Climates, midsummer and midwinter are commonly the most healthy parts of the year, and on the other hand epidemic diseases most frequently prevail in the spring and fall of summer. It is highly probable that we are defended from some species of putrefaction in summer by the vegetables forming an immense quantity of pure or dephlogisticated air.

But during the heats of spring and autumn these diseases prevail in the greatest degree, and it is natural to be expected especially in the Latter season when putrefaction prevails wherever heat sufficient occurs.

Even the plague itself seems to depend almost entirely on season. This remarkable disease was totally unknown to the Ancients. Neither Celsus nor Galen mention it, but the first time we find to have occurr'd was in the days of Justinian, and it is first describ'd in the writings of Procopius.

It depends entirely on temperature and is destroy'd either by great heat or cold. It has been doubted whether it does not rise from the fermentation of putrid miasmata, or from animalcula, but in both these cases the agency of heat must be very considerable.

We come now to make some remarks on

CLIMATE

It is an obvious but very interesting point, how much every Climate corresponds to the human frame!

Mankind can inhabit this globe perhaps from the pole to the line for altho it has been as yet inaccessible, late voyagers have penetrated within 8 or 9 degrees of it and found it perfectly inhabitable, so that very likely the human species can inhabit this globe from pole to pole. But there is no other organiz'd body to which this observation applies, neither plant nor animal.

We find however that Climate has a very powerful effect upon all organiz'd bodies, upon Vegetables these appear every where. The same species of plant is often found in different climates both in the old and new world, in these different situations however there are considerable variations superinduced by climate. The same likewise takes place with respect to all animals. The same species is often found in the old and new world, but varies so considerably in form and sometimes in manners that they have often been mistaken for different species.

We know the effects of Climate likewise on the human frame. The external characters of mankind are exceedingly different according to Climate, and we know that these external characters which have subsisted for many ages can be altered by transpositon of some generations into a different climate.

There is a curious question with respect to the Comparative nature with respect to Climate of the Northern and Southern hemispheres.

We would be apt to imagine that these in similar Latitudes were merely the same; but from facts we now find the case to be quite different. The ingenious Dr. Martine[82] in the year 1740 in his treatise on heat, in consequence of his careful researches was as far as I can find the first who noted

---

[82] Benjamin Martin (1704–82), English mathematician and optician, *Natural History of England* (London, 1759–63).

that there was a remarkable difference between them, and that similar Latitudes of the Southern hemisphere were colder than those of the Northern.

This opinion however has been lately combated by Buffon who affirms that the Southern hemisphere is equally warm with the North.

I shall now enumerate those instances which throw sufficient light on the subject to shew that Buffon's opinion was a mistake; and first I shall mention these authenticated facts with regard to the Southern hemisphere which we owe chiefly to the late circumnavigators.

Fifty or Sixty years ago a Captain Tunnell, who sailed around the world and returned by the North of Scotland, found at Cape Horn in Lat. 60° S. in the middle of summer a colder and more boisterous Climate than Lat. 62° N. on the north coast of Scotland, in the midst of winter.

The next accounts of this matter we had from Admiral Byron who in his voyage round the world said that he experienced worse weather in 45° S. Lat. in the midst of summer than he had ever met north of the Bay of Biscay during the depth of winter.

Prince Edwards Isles 46° S. Lat. were found by Captain Cook during the Summer completely covered with Snow.

In the Isles of Desolation in the middle of December he observed the Thermometer stand always about 34° or 36° at midday.

In December 11th 1763 Capt. Cook encountered the first ice in S. Lat. 61°, and did not lose sight of it till in S. Lat. 51° and he could no where penetrate the Southern hemisphere farther than till 71° S. Lat. In the Lat. of 51° S. he saw Islands of Ice. In the Lat. 49° S. he found the freezing degree on December 9th and in S. Lat. 50° encountered Ice that same day.

We shall now compare these with the observations that have been made in the Northern hemisphere.

In Captain Cook's voyage to the North he did not fall in with ice till in Lat. 67°, and upon another occasion when sailing between Asia and the West Coast of America he did not see Ice till in 70° of Latitude. We had an instance a number of years ago of a Greenland ship which sailed from Leith in 1751 that penetrated nearer the pole than any one had done before or has been able to do since. She sailed up Davis's straits till 93° and a half and she did this without being in the least interrupted by Ice. And the Captain of another Greenland vessel said he never found the Ice troublesome farther South than 74° N. Lat.

It appears therefore upon the whole that the Southern hemisphere is undoubtedly much colder than the northern.

The cause however of this difference is far from being so obvious.

It may however be admitted as some part of the cause that the Earth's orbit is nearer the Sun in the Arctic winter than in the Antarctic, and the Arctic Solstice is 10 days longer than the Antarctic.

But the chief reason is the great extent of Land on the Northern hemisphere, whereas in the Southern hemisphere there is no considerable body of Land in the higher Latitudes, but is almost entirely an immense waste of waters, while those on the Northern side of the line are mostly occupied by land, and we know that the suns heat is absorbed by deep water while it is reflected by the land.

Exclusive of the Barometer it has been customary for all nations to form Prognostics concerning the future state of the Weather from various other appearances, as from particular appearances in the heavenly bodies, the Sun, moon, Stars etc. from different states of the Atmosphere, from inanimate substances, from plants, from Animals and even from Man himself.

Thus the paleness in the Suns body has always been considered as portending rain, and redness fair weather.

With respect to certain Animals they have probably a degree of knowledge concerning the future state of the weather with which we are entirely unacquainted.

It has been now the practice for more than 150 years past the keeping

REGISTERS *of the* WEATHER

We generally in these find a column allotted to the Barometer and another to the Thermometer.

But the great deficiency in these is the wanting those observations relative to the state of the weather which one would think every one would make.

I have here the form of a Register which appears to me the best yet invented.

It consists of Seven columns. The Ist contains the Days of the Month. The IId Observations on the Barometer at noon. Mr. De Luc's Barometer is the one used with Nonius' divisor, by means of which the fractions of an Inch can be measured to 1/500th part and with a good eye even to 1/1000th part. The IIId and IVth Columns are allotted to the Thermometer. It has been the common practise with some to observe the Thermometer at Midday, and with others at 8 O'Clock in the morning, and where the Thermometer is only to be noticed once in the day, 8 O'Clock in the morning is the best time of the two, as being nearest the medium temperature of the day.

But here the IIId Column marks the state of the Thermometer at noon and the IVth at Midnight; and in order to have an accurate view of the temperature of any particular place, this is necessary.

The Vth Column denotes the days which are fair or those in which there has been snow or rain.

The VIth column is appropriated to the Winds and here all that we can do is mark their directions.

The VIIth is allotted to accidental remarks which may have occurr'd.

But in this diary at the conclusion of each month there is likewise added such observations with respect to the weather which may have occurred to the observer. The want of such observations seems to be the principal deficiency in most diarys that are kept of the weather.

We come now to another branch of our subject vizt.

HYDROGRAPHY

OR

The History of the Waters of the Earth

And here it will be proper first of all to take a view of the different kinds of Waters their

DIFFERENT DEGREES *of* PURITY

and their different properties.[1]

The purest water with which we are acquainted is Distilled Water. It is much purer than any natural water for the purest water which nature affords always manifests a quantity of terrine particles.

Mr. Boyle long ago distilled the same Rain 200 times and constantly observed a small portion of a white earthy sediment but this was so small in the latter distillations that it was nothing but a small whitish speck in the glass retort.

---

[1] The chapter on hydrography is superb for the 1780's. Knowledge of water, and especially of groundwater, was still somewhat involved with witchcraft. Walker had an unusually good understanding of the origin and movement of groundwater. He made many observations on this matter in more than thirty visits to mines and by the study of springs and running water. His first scientific paper (1758) was concerned with the composition of water from the subsurface. Almost everything in this chapter is now included in hydrogeology.

Hoffman[2] was likewise very well acquainted with this impurity of water, but he imagined that all the terrestrial matter was merely accidental, and no constituent part of the water itself; and indeed since his time this has been the general opinion, till of late years that some philosophers have maintained that even the purest water was wholly convertible into earth by means of repeated distillations, and even that this might be effected by trituration. But these opinions however ingenious remain to be confirmed by unequivocal and incontestable experiments.

Water distilled even once is purer than any natural water and yet after frequent distillations have made it extremely pure, when used as drink or with diet it always gives a sense of weight at the Stomach; this we may naturally presume owing to its being depriv'd of the Air it before contained. Hence Mineral waters, notwithstanding the mineral impregnations with which they are often very considerably loaded, are the lightest on the Stomach on account of the large portion of air they contain.

The purest natural water is that obtained from Hail; on distillation it gives the least residuum of any natural water and indeed it is natural to suspect this from its being formed in the most elevated parts of the aqueous exhalations, where heavy and terrestrial particles never reach, in consequence of this it is the purest of natural waters. In winter it is often so pure as not to contain above one grain of terrine matter in an English Gallon.

Next in purity to this is

Snow Water

It is sometimes so pure that it has only four grains of a residuum in an English Gallon, but at other times it is much more impure. It is however never liable to that degree of putrefaction as rain water.

The next in purity is

Rain Water

It has been more frequently subjected to distillation than other species of water, and it has generally been found to contain in an English Gallon from 6 to 12 grains of an earthy residuum. It has always more sediment in summer than winter, and in a hot than in a cold climate.

The purest Rain water is always liable to putrefaction which proceeds from the Subtile Animal and Vegetable matters it is impregnated with, for on distillation it exhibits not only an earthy residuum, but likewise a small portion of oily and inflammable matter.

[2] Friedrich Hoffmann, chemist, University of Jena, born 1660.

The terrestrial matter obtained from Rain water was most accurately examined by —— he found it to be a subtile white calcareous earth. From 100 measures of rain water of oz. XXXVI each he obtained grs. 100 of this earth; but from an equal quantity of Snow water only 60 grs.

The water of

SOFT SPRINGS

is the next in purity to Rain water.

These vary considerably as to their contents; but in general our purest perennial springs contain from 5 to 40 grains of earthy residuum in an English gallon.

Pure Lake water contains from 4 to 40 grains of residuum in the English gallon.

The Water of Dew is more heavily loaded with terrine matter than any of those already mentioned. It has been found to contain from 18 to 140 grains of earthy residuum in an English gallon. It seems to be in consequence of this that it is much more liable to putrefaction than the other waters before enumerated.

PIT WATER *or as it is called* HARD WATER

contains sometimes a very great quantity of earthy matter for from 40 to 316 grains have been found of earth and saline matter in the English gallon.

These pit springs will not make a lather with Soap; the reason is that they always either contain an earthy or metallic salt, and hence they decompose the soap by the acid in the saline matter attracting the Alkali of the soap and leaving its oleaginous part disengaged.

These hard waters however by having something of more taste than the purer waters which are always very insipid, are generally esteemed better to drink.

They refuse to make a proper infusion or (as we commonly speak) do not draw tea, nor boil vegetable substances as peas etc. so soft as other water, but they preserve the green colour of all Vegetables that are boiled in them, much better than soft water.

It is well known that our hard pit waters are rendered quite soft by exposure, and this evidently happens from their depositing their fossil contents.[3]

[3] The hardness of water is a complex subject, and although Walker does not have a complete understanding of it, his knowledge is remarkable for the late eighteenth century.

It is likewise known, but which is not so easily accounted for, that wells of soft water excluded from the air have become hard as with a close pump.

With respect to

RIVER WATER

It is exceedingly irregular in its degree of purity; sometimes it is equally pure with spring waters when the river is composed of many spring waters. At other times especially in floods it is heavily loaded with earthy matter.

I may here observe in general that the purest water is always the freest from earthy matter, and hence it may be reasonably concluded that the terrine particles found in water are merely extraneous.

Here I may take notice of an effect commonly ascribed to the use of Snow water the truth of which however is very dubious, and that is the famous disease known by the name of *Bronchoccle* or *Hernia gutturis*.

It is a tumour or wen growing on the neck; but is certainly very improperly termed Bronchoccle as it is very plain it has nothing of the nature of Hernia.

It prevails in many tracts of Switzerland particularly in the province of Allais [Valais].

The tumour or wen itself is from the size of a walnut to that of a man's head. The people subject to it have always an impediment in their speech. In a village near —— in Allais almost the whole inhabitants are affected with it, some of them are born with it and others have it from 8 or 10 years of age. By all accounts it appears that it is to a certain degree hereditary, hence it generally descends to the children of goiterous persons, sometimes however the children are entirely free of it, but this is well known to happen to other hereditary diseases particularly the Scrophula.

Scalinger[4] was the first who attributed this disease to the use of Snow water, and even now the general opinion is that it arises from the Snow water containing a great deal of terrestrial matter, but this we find is by no means the case.

Dr. Forster on board Capt. Cook observed that when the sailors were obliged to use Ice water in the high Latitudes, it produced swellings in the Throat. This he attributed to the want of fix'd air which he supposed is expelled by freezing. We are assured indeed by the experiments of Bergman that Snow and Ice water have little fix'd air in them.

But we must have recourse to some other cause for this disease than the use of Snow water, for there are many other inhabitants of Alpine countries and

[4] Joseph Justus Scaliger (1540–1609), professor in Calvin's college at Geneva, 1572–74, son of Julius Caesar Scaliger (1484–1558).

even in other parts of Switzerland who make use of equally as much or even more Snow water that are quite free from this disease.

It appears to be merely a local disease and exists not in those parts of the Alps where Snow water is most used, and it is known that it can even be prevented in very young children by sending them to the mountains.

We have lately got an account of it as subsisting in the Island of Sumatra and it is particularly described by Mr. Marsden[5] who gives a very interesting and distinct account of the Island, and there where snow is not known this very goiterous disease does likewise exist.

Mr. Marsden thinks that it is owing to foggy air, but this cannot be the case as in much more marshy countries this disease does not subsist.

There is however one circumstance in Sumatra which corresponds with the goiterous part of Switzerland, and that is in both countries the waters are stagnant and their channels are heavily loaded with putrefying and terrine matter, and accordingly in the channels of the Rivers, in that part of Switzerland where the disease prevails, are lined with a real Tupha or calcareous crust.

This I consider as the most probable cause of this disease which it may produce by occasioning glandular obstructions.

Another melancholy disease prevails likewise in Switzerland in the same country with the Bronchoccle, and is commonly ascribed to the same cause. Those parts of Switzerland inhabited by the goiters contain more of those people than any other place. Those afflicted with it are called Cretins and the disease itself La Cretinache. The people labouring under this disease are always somewhat dull of hearing and have to some degree an impediment in their speech, they are frequently both deaf and dumb and labour under nervous degrees of imbecility both of mind and body, and even a degree of insanity, and are frequently likewise goiterous.

At G——— Count Razoumousky[6] a Polish gentleman found no less than 60 persons labouring under this disease in one small parish so that there were Cretinous persons in almost every family.

The Cretins are of a paler complexion than the other inhabitants, they are a feeble race both in body and mind, they are very easily irritated, liable to chagrin, pusillanimous, and usually incapable of energy unless by giving way to a puerile passion. Human nature here subsists in a very degraded state.

They are treated by the more enlightened sort of people with humanity and pity and by the vulgar with a superstitious regard.

Both the goiterous and cretinous persons marry among themselves, with

[5] William Marsden (1754–1836), *History of the Island of Sumatra* (1783).

[6] G. de Razoumovsky, *Voy. Min. Phy. Bruxelles* (Lausanne, 1783).

each other, and with the other inhabitants of the country by which means the disease is strengthened and propagated.

The whole inhabitants of these parts of Switzerland are melancholy, labour under an embarassed respiration and are shortlived, and a degree of this malady seems to pervade the whole.

Count Razoumousky ascribes the disease to mephitic air arising from the marshy grounds, but this can not be the cause.

But a body of people constantly using tophaceous water is certainly a rare and peculiar occurrence and it may reasonably be presumed to produce some rare and peculiar disorders. It is not improbable that both diseases (the Bronchoccle and Cretinache) are owing to the same cause, and that these different appearances are occasioned by the mere direction of the disease from the throat to the brain.

It would be worth while to detect whether Cretinous persons are born insane or whether they only are this by length of time. The dissection of their Brain has not as far as I know yet been attempted. But such a dissection of these persons at different ages would very probably lay open some very considerable discovery.

Here I may take notice of a singular phenomenon the

LEVELLING *the surface of water by means of* OIL!

Oil thrown on the surface of the stormy sea immediately calms the tempest and renders the water pure and pellucid. This fact was well known to the ancients and is particularly described both by Pliny and Plutarch, and it was practiced by the ancient Coral fishers in the Mediterranean, and it is yet used by those people when they want to see to a great depth, for the smallest rusling of the water is sufficient to obstruct the view, but when they throw a little oil on the surface the water becomes as smooth as a mirror.

This is likewise a practice that may be of great use to naturalists when employ'd in acquiring and examining the marine productions, and it was used with success by Sig. Donati[7] when he collected the submarine productions of the Adriatic.

I have likewise frequently practiced it myself. There is one thing however that when the oil is poured from the Boat it has not the desired effect till it comes in contact with the side of the Boat, and then you see distinctly for four or five fathoms depth, whereas before you could not discern a foot depth.

The fishers in Spain too when catching oysters and other shell fishes use

[7] V. Donati, *Essai . . . Adriatique* (La Haye, 1758).

this method for keeping the Sea smooth, and when the oil is poured upon the water it dilates to a great extent and becomes a most surprizing degree of firmness, and when this is properly done a person may get a distinct view to the depth of 8 or 10 fathoms.

We know that the repulsion between water and oil is very great, and this is applicable to the making of paper, when they are troubled with an aqueous froth they pour some oil on it by which it is immediately dispelled; when it is wanted likewise to check the progress of fermentation in wine they throw a vine leaf into the vat with some butter or oil on it which immediately checks the fermentation.

Of late years the celebrated Dr. Franklin made many ingenious experiments on this subject. The Dr. who was not displeased to be thought acquainted with magic used some times to go to the sea shore and when stormy command it to be calm, striking it with his cane which was hollow and full of oil. He carry'd it so far with regard to oil that he thought it might be of use to ships at sea by smoothing the surface of the ocean in a storm, that even a few casks might prevent shipwrecks; or might save the lives of the ships company of the wreck.

A German author of Göttingen takes particular notice of this and calls the Dr. *"Vir audacis ingenii"* who not content of depriving Jupiter of his thunder bolts now attempted to defraud Neptune of his Trident.

The Cause of this effect of Oil in smoothing the surface of water is far from being obvious, besides its repulsion for water oil may likewise have a strong repulsion for Air.

We come now to make some observations on the

NATURAL HISTORY *of the* OCEAN

Where the first thing that occurs is with respect to its Level.

It is commonly supposed that in consequence of the general law of gravitation in fluids the ocean behoved to be every where of the same level. Of late however this has been attempted to be controverted.

Mr. Culm,[8] the secretary of Dantzic, publish[ed] some time ago a treatise on this subject, in which he throws out some extraordinary ideas, and ascribes such inequalities to the level of the ocean in different places as can by no means be admitted. They are extraordinary, theoretical and without any support.

We shall examine only one instance. By a laborious sort of computation he

[8] Perhaps Johann Adam Kulmus (1689–1745), professor of medicine and natural science at the Gymnasium of Danzig.

pretends to find that the level of the Pacific Ocean to the westward of the Isthmus of Darien in America is no less than 14.900 feet higher than it is on the Gulph of Florida on the east side of that Isthmus.

Now with the nature and situation of this Isthmus we are pretty well acquainted, from which this opinion must appear utterly impossible.[9]

I have observed something like it in our own country in the Isthmus of the promontory of Kintyre. That Isthmus is about 40 or 50 miles long and in some places only 3/4 of a mile broad, one thing is certain that at some particular times the sea is from 10 to 20 feet higher on the West than on the East side, and vice versa, but this happens because tides make at different times on the opposite sides.

No doubt the Level of the sea is in our high Northern Latitudes a little higher in winter than in summer, at least in all the narrow seas, which arises evidently from the winter rains increasing the river which discharge themselves into these seas and from the diminished evaporation of the season. But how far this exists in warmer countries is not so far as I know precisely determined. In the Adriatic however the level was found by Sigr. Donati 2 Roman feet higher in winter than in summer, this must be owing likewise to the casual inundations to which it is subjected.

We shall next attend to the

Access *and* Recess *of the* Ocean

or its encroachments on some parts of the earth and retreat from others.

On this subject there is a curious opinion which was only entertained by Linnæus and which he has endeavoured to corroborate in his writings, and that is that the whole of the inhabitable globe was formed gradually out of water from time to time deposited in a solid form while the ocean receded. This was a very ancient opinion and we find it clearly mentioned by Aristotle that the ocean has gradually deserted the land.

It appears not however that as far as history or tradition goes that either the access or recess of the ocean have been considerable.

Let us take notice of Italy which was the source of ancient history. We find on the northern shores of Italy in the Bay of Boya houses and streets are distinctly seen under water to this day. Near ——— the remains of houses, gates and windows are likewise to be seen under water. In the year 1664 on the coast of Tuscany, Kircher saw the ruins of a whole town under water.

[9] Walker was correct in questioning the reported differences in the level of the sea. He correctly ascribes these differences to the effect of the tides and especially to the different times of the arrival of the tides along the coasts.

In like manner as the Sea has made encroachments on the Western shores so has it receded from the Eastern coast. Thus the port of Bernice which in the consular times was a principal harbour and contained ships of a very great size, can now contain nothing larger than a fishing boat. The sea has likewise receded from the port of Rimini 1750 English feet, and at Ravenna which in the consular times was a considerable port. The seat of that city stands three Italian miles from the sea shore and all the intervening space which 1700 years ago was occupied by the Sea consists now of cornfields, vineyards and gardens.

There was a remarkable fact at Venice discovered in the year 1722 when they were about to form a new pavement in the place of St. Mark. In digging they found another pavement 5 feet below the old one which was 2 Roman feet above the level of the Sea, so that it appears the sea had risen in this place 3 feet.

If such has been the case in the Mediterranean where the tides are exceedingly small and where the sea is a mere mill pond to the ocean; what must it have been in other parts of the globe where there are much higher tides and where heavier seas prevail.

In Britain we have many instances of the encroachments of the Sea on some parts while it has deserted others; of its encroachments the historical account of Earl Goodwins estate, now the Goodwin sands,[10] is a very remarkable fact. It has gained considerably on the East coast of Britain but its encroachments are chiefly confined to those parts where the shores are soft, land low, or where there are high earthy cliffs, but where they consist of fix'd rock there is little either of diminution or increase.

In like manner in Holland tho' the industry of these republicans has been supposed to have barred out the sea yet on the Dutch coast the ruins of a Roman fort are discoverable in the middle of the water which was 3 miles from shore. The foundation of which at low water is now 10 or 12 feet sunk, so that in Holland as in our country what the sea gains in one part it seems to lose on the other.

There have been at times in every quarter of the globe violent and accidental inundations of the ocean owing to the tides and winds as in some parts of India and North America, and when the memory of these is gone, the exuviae that remain of them induce people to think that such parts were steadily occupied by the sea which perhaps only felt its effects for a single tide. Many of these instances occur on our own shores.

[10] William Lambard, in his *Perambulation of Kent* (1570), states that according to an old Saxon chronicle the Goodwin sands came into existence in 1099. They are sands along a tidal flat that still form quicksand and barren stretches of desolate land.

Mr. Stellar[11] remarked on Byrons Island between Kamschatka and the American continent, in one place for a great extent, great quantities of drift weed and the Skeletons of Animals thrown up above 30 fathoms above the level of the sea.

We have no authentic instances to show that the sea does any where subside into the bowels of the earth.[12] We find no where the waters of the ocean in the interior parts of the globe; on the contrary we have many instances to show that a very thin crust is capable of supporting the waters of the ocean and preventing their subsidence. At Whitehaven in Cumberland they work the coal for a measured mile under the sea bottom. The West India ships sail over the heads of the Colliers and yet there is no sea water found in the mines. At Borrowstonness I saw at one time above 2 acres of Coal wastes as they are called about 20 fathoms under the bottom of the sea and yet the whole water was carryed off by the labour of a single man.

At Kinnell in that neighbourhood they dig a quarter of a mile under the water and in one place to within 12 fathoms of the bottom and yet these are among the dryest mines in the country.

In Cornwall they frequently drive a tin mine till within 8 feet of the bottom of the sea and yet such mines are often dryer than those in the interior parts of the country.

One thing appears in general that the balance of terrestrial matter is rather in favour of the ocean by the terrestrial matter carry'd off in the beds of rivers and deposited in the bottom of the sea. But so far as tradition goes, the ocean has neither increased nor diminished to such a degree as to allow the constitution of the habitable globe, and in short my opinion is that Caves obtain none of Neptunes dominions without paying an adequate compensation.[13]

We come now to consider the

Saltness *of the* Ocean

On this subject it has been a much controversial point whether the saltness of the sea was coeval with the globe? or whether it is only the work of time? and indeed the question is not of easy solution.

[11] G. W. Steller, *Besch. Kamachatka* (Frankfort, 1774).

[12] A popular theory, which is why Walker takes pains to point out the fallacy of the concept.

[13] Walker has documented well the encroachments of the sea upon the land as well as the retreats. He recognizes a balance between land and sea but fails to appreciate the extent and periodicity of sea invasions.

The Aristotelians and the Cartesians were all of the latter opinion and indeed it has been the general opinion that the sea was formed fresh.

It was the opinion of Kircher[14] that the salt was furnished by subterranean fire and communicated by subterranean canals. Others with more probability have thought it derived from submarine rock salt.

The celebrated Dr. Halley formed a sort of Theory on this subject, and supposed that the ocean in course of time obtained all its salt from the Earth by means of Rivers. But River water contains only 1/4000 part of Sea salt, and hence all the rivers in the world would require 6400 years to give to the ocean 1/500 part, but we know that the waters of the Sea contain in many cases considerably more than 1/30 part of salt.[15]

There is one thing certain that since observations have been made on the subject which is now a considerable time, no fact has been brought to prove that the Saltness of the ocean has increased in any degree. It may likewise be observed that myriads of plants and Animals which it contains could not subsist in fresh water and also that they would be hardly able to subsist were the saltness greater than it is.

It may likewise be observed that great inland Seas as the Caspian which receives many rivers and has no communication with the ocean continues perfectly fresh.[16]

There is another argument which arises from the putrefaction which would ensue in the inlets and arms of the sea if the waters were sweet and in the warmer parts of the globe the sea would become a huge mass of putrefaction.

These reasons seem to all point at this, that the water of the ocean in respect to Saltness is pretty much at present what it ever has been.[17]

We shall next consider the

[14] A. Kircher, *Mundus subterraneus*.

[15] That Walker recognizes variation in the salinity of the ocean is not the most interesting part of this paragraph. Our attention is focused on his attempt in passing to determine the age of the ocean by calculating the rate at which the rivers of the earth add salt to the oceanic basins. This is one of the earliest of such attempts.

[16] We can hardly believe that Walker intended to use the Caspian Sea as an example of a fresh water body. A few pages further on, when discussing inland seas, he describes the salinity of the Caspian in considerable detail and remarks that it is higher than that of the Mediterranean.

[17] This conclusion is one of Walker's finest. W. W. Rubey in a paper on the composition of the ocean (*Geol. Soc. Amer.* 62 (1951): 1111) concluded that "the composition of both sea water and atmosphere may have varied somewhat during the past; but the geologic record indicates that these variations have probably been within relatively narrow limits. A primary problem is how conditions could have remained so nearly constant for so long."

Degree of Saltness in the ocean and the contents of Sea water. Dr. Lucas[18] found at Harwick that the sea water contained 1/25th part of Solid content. At that place an English pint of Sea water was found to contain of Solid matter 4 drams, 2 Scruples and 19 grains, in this solid residuum were 4 drams of pure sea salt, 30 or 40 grains of bitter salt part Magnesia muriata and part Magnesia vitriolata; 8 or 10 grains of calcareous earth, and the remainder an oily matter.

This is to be presumed as the most common contents of the sea water, but the proportion of these varies considerably in different climates.

The Saltness of the ocean is far greater in the warmer climates than in cold. We are assured that the sea water under the line has the quantity of 1/17th of solid contents of which the far greater part was pure salt. This greater degree of saltness in the water of hot than of cold climates seems to be a happy precaution of nature to balance the greater tendency to putrefaction in the former.

The waters of the Mediterranean are salter than any part of the ocean under the same latitude. But in opposition to this the Baltic is not so salt as the German ocean. In the British it is from 1/27 to 1/30 of solid contents. Dr. Lucas found the water of the Baltic to contain about 1/30th only.

By experiments made at the saltworks in Fife near St. Andrew's in the Firth of Tay, it was found that the sea water on that coast contains 1/30th part of solid contents, and 1/40 of pure sea salt. In different parts of the Firth it is found to contain 1/29 or 1/28 of solid residuum, and in that part of it where the sea is saltest in the bay of Kirkaldie and Dysart, 1/27th.

Upon the whole the Saltness of the ocean appears to diminish gradually from the line to the poles. In the torrid Zones it is always the saltest. There it was found by Dr. Lucas to contain 1/7 of solid contents the greater part of which is sea salt.

The ingenious Capt. Ellis performed many experiments at Sea on this subject. In one of his voyages he was furnished by Dr. Hales with what he called a bucket seagage, a machine by which he could draw out of the sea, water at any determinate depth. By the help of this instrument he found that at any considerable depth the sea water was always heavier than at the surface which he was assured of both by weighing it against fresh water and with an aequipoise of sea water taken up near the surface.

With regard to the Nauseous taste of Sea water. It was first found by Bergman to be confined merely to the superficial parts of the ocean, and yet it was this taste that for a long time prevented the use of Sea water after distillation because it always adhered to it after distillation.

[18] Unidentified.

The Distillation *of* Sea Water

Has always been a great problem. To attain it many experiments have been made that evidently either fell short of the end proposed or could not be made universally useful. The great desideratum was to prevent the distilled water from retaining the nauseous taste.

The first experiments on this subject were made by Capt. Chapman in 1757. He found the best way of distilling sea water was by throwing wood ashes into the still which probably was nothing more than adding to the water a small proportion of fix'd Alkali. By this mode of distillation a good potable liquor was obtained from sea water, and he says that it was known to Sir Richard Hawkins[19] tho' he described it rather in an enigmatical manner for he says that he could distill a hogshead of sea water by means of Billets of wood which Capt. Chapman says is impossible except by their ashes.

Capt. Wallis was the first who made any considerable use of it, in his voyage round the world. From 56 gallons of sea water he got by distillation 36 gallons of pure sweet potable water. It was his custom as soon as his stock of fresh water was exhausted to 40 tons, to set his still.

Of the

Inland Seas

we have only one to take notice of—the Red Sea. This is certainly a subject that affords a world of curiosity. But unfortunately to the history of this remarkable sea and its shores we are as yet very great strangers.

The Red Sea has evidently its name from the Red sand which every where abounds on its shores, and this I suppose proceeds from the Red granite, which by all accounts is the prevailing rock in that sea and on its shores.[20]

This sea is likewise remarkable for the production of Red coral, and perhaps this too might have given rise to the name.

Here we may take notice in general of the degrees of Saltness, and the tides of these Inland Seas.

It appears to be a mistake that the Mediterranean is salter than the ocean. Its waters appear to be much the same, or if there is any difference, rather less salt than that of the ocean in the same Latitude.

The waters of the Euxine Sea and of the Palus Meotis[21] are entirely fresh.

---

[19] Sir Richard Hawkins (1562?–1622), naval commander.

[20] The theory that the Red Sea derived its name from the red sands along the shore may be as good as any, but the name is commonly ascribed to the red pigment in a blue-green algae, *Trichodesmium erythraum,* which occasionally colors the surface.

[21] Sea of Azov.

The Caspian Sea differs considerably; though it has no communication with the ocean yet it is considerably salter than the Mediterranean. This appears to be owing to the nature of the adjacent countries which are remarkably replete with Sal Gem, and likewise abound with the fossil Alkali or Naptha, and it is probably from the abundance of these salts that the waters of the Caspian are endowed with such a remarkable degree of saltness. There is one thing pretty curious in the history of this sea and which I shall have occasion to mention afterwards, that in consequence of this saltness it is inhabited by a great number of marine Animals, which is not the case with any of the inland seas, or with Lakes, the waters of which are fresh.

But the most remarkable inland sea with regard to saltness is the Mare mortuum or Dead Sea of Judaea the waters of which are so heavily loaded that the saline matter contained in them amounts to about oz. VI in a Lb. of water, of this quantity of saline matter however only about oz. I is pure sea salt the other oz. V being earthy salts, some of which have calcareous earth for their base and other Magnesia.

The waters of the Baltic are somewhat fresher than those of the ocean.

As to the

Tides

of these Inland seas.

The Mediterranean has a small degree of tide content by depending on the attraction of the Moon. But in general the ordinary tides are little more than one foot. They increase however regularly at the fall and change of the moon, and in some places of it at the solstices the tide rises to no less than $5\frac{1}{2}$ feet.

In the Thracian Bosphurus, at the straits of Constantinople, there is a vicissitude of tide of about 8 or 10 minutes.

We are told that there is no perceptible tide in the Euxine sea, Palus meotis or in the Caspian Sea, but this is rather somewhat to be doubted, for such an immense body of water, particularly the Caspian, must be in some degree or other affected by the Moon. Accordingly Lake C[hamplain], in N. America though far inferior in size to these seas has a vicissitude of ebb and flow depending on the periodical returns of the Moon, and we are informed by the French that Lake Superior the greatest body of water in America has a tide and ebb at full moon amounting to no less than 3 or 4 feet.[22] If this be true

---

[22] I do not know the source of this figure, but it is false. The tide is of the order of one to three inches. It is rather characteristic of Walker to follow such a statement with "If this be true. . . ." Some of his information about American lakes probably came from J. Carver, *Travels through the Interior Parts of North America in the Years 1766, 1767, and 1768* (London, 1778).

we can have little doubt that the Caspian is subjected to the vicissitudes of tides.

The Baltic has a small degree of flux evidently depending entirely on the moon and not on its communication with the ocean, for when swelled by the Autumnal rains it sends forth a current into the German ocean, and in the heat of summer when it recedes it receives a current from that sea.

We come now[23] to treat of the

LUMINOUS *or* PHOSPHORESCENT PROPERTY *of the* OCEAN

In treating of this subject, I shall give an account of the different circumstances attending the Phenomena, and then of the theories which have been proposed to account for it.

Everyone is acquainted with this property, who has even been at sea. Though it is one of the most common Phenomena of the Ocean, yet no one to this day has fully described it. Those who have endeavoured to assign the cause of it, have generally had little opportunity to examine the Phenomena itself. In the Ocean there are great quantities of phosphorescent substances and animals; but these seem not capable of explaining it.

Linnæus ascribes its shining to the *Nereis noctiluca,* one of the sea vermes, which possesses a phosphorescent quality.

Query. Whether Linnæus means this appearance? The sea sometimes appears illuminated for a considerable extent, which is undoubtedly owing to insects, as they have been discovered by the microscope. Vid: Cook's Voyage 29th October, 1772.

The Abbey Nollet[24] likewise ascribes this appearance to sea insects. Ferber[25] the Swede has lately attributed it to vast quantities of Medusa and Moluscas, that are everywhere found in the ocean.

Canton[26] of the Royal Society, made many experiments to prove that it is putrefied animal substance incorporated with the sea water.

Some have ascribed it to an emission of the light absorbed by the waters from the sun during the day, and Dr. Franklin, when he first paid attention

---

[23] The following sections: (*1*) Luminous . . . Property of the Ocean, (*2*) Depth of the Ocean, (*3*) Heat of the Ocean, (*4*) Water Spouts, (*5*) Typhoons, (*6*) Whirlpools, and (*7*) Tides, are from D.C. 2-18. There is very little difference in wording between the older and the revised versions, except that in a few instances sections seem to have been omitted from D.C. 2-33. The apparent omission may have been due to faulty copying by a student or a professional copyist. The lectures are presented here in as complete a version as possible.

[24] Jean-Antoine Nollet (1700–70), *Essai sur l'électricité des corps* (Paris, 1750).

[25] Johann Ferber, *Travels through Italy.*

[26] John Canton, F.R.S., English physicist.

to this phenomena, attributed it to electricity. He afterwards retracted this opinion; but I still think it most probable that it arises from electricity, especially on account of the dependence it has on the state of the atmosphere.

I had occasion for 8 or 10 dark stormy nights at sea to mark the circumstances of this phenomenon with great care. I had a long paper of observations relating to it, which, I believe I have now lost; so that I can only give you these observations from what I can recollect. This luminous property of the Ocean, in our climates, occurs more in autumn, spring and winter than in summer. There are certain nights of the darkest kind, in which no light at all can be excited. I remember that on these nights in which the sea was most luminous, the Barometer was always extremely low, and the air in a moist state. Hence the Sailors judge it the most certain sign of foul weather. On the contrary, when the Barometer was high, though the night was ever so dark, no light could be excited in the sea water. This observation seems to point at the cause of this Phenomena, in so much that it seems to depend greatly on the state of the atmosphere.

As to the appearance of the light, it appears in quick and successive flashes in the place where the water is struck by a solid body. The light is extremely agreeable to the eye, being too feeble to offend it by its splendour, and its colour is a tinge between blue and green, but of so dubious and indeterminate a nature that to different Spectators it seems either. The light it most resembles in splendour and colour, is that of the glow worm.

I observed, that the quantity of light produced, was in proportion to the force applied and to the quantity of water there agitated. For example, a much greater quantity of light was excited by a smart stroke of a cane than when the cane was applied softly; and that the motion of an oar in the water, tho ever so gentle, excites a much greater quantity of light than could be produced by a walking cane, though most forcibly applied.

There is another remarkable circumstance attending this appearance, which is, that is always produced in flashes of a spherical figure, or of a figure approaching to a sphere, in large apparently round luminous bubbles from the size of a pea to the size of an orange. When a vessel is under sail in the night, these bubbles, or light, are produced at the bow and along the sides of the vessel in great quantities; from whence they are carried into the wake, and then the light appears at its greatest perfections. At the stern you can see the bubbles distinctly, but when they get into the wake, they coalesce and continue for about the distance of the length of the vessel and then disappear. The light is so strong that a person might read a book of large print at the stern of the vessel, or observe the hour or the minutes on his watch.

There is another observation which I made, that the light is only excited by a force applied with a hard body and that the light is produced precisely in the place where the surface of the sea and the surface of the hard body meet; for I have seen the waves rolling on each side of the ship at a little distance, and agitated violently, yet I never saw them afford any light; while at the same time the water touching the sides of the ship, and the wake appeared all in a flame. When a boat hook too was pushed down, the light proceeded from the surface of the pole, and as the bubbles were observed to rise along the pole, it would appear that they have a degree of attraction to a hard body, for they run up the pole with a perceptible degree of cohesion.

I examined the salt water itself and this time with great care when this appearance was in great perfection. It was at a great distance from any land, where the sea was of the depth of at least 100 fathoms; and on inspecting the water, if any water could be said to be purer than the purest spring water, it was this. I examined it by a strong light; and though I employed very great magnifiers, I could not find a particle of heterogeneous matter in it.[27] The light therefore does not proceed from marine insects or putrefied animal substances, but appears to be altogether a *sui generis* light, and most probably of an electrical nature; and by no means are we to agree with Mr. Lyon, that this light is owing to an emission of the solar light from the water.

We have but few experiments made upon the

Depth *of the* Ocean *at* Different Places

The Gentleman whom I formerly mentioned, Captain Willis, sounded to the greatest depth I have ever heard of in making some experiments. He let down his Bucket sea-gage to the depth of 5346 feet, which is about 60 feet more than an English mile, without finding bottom. This was in the Atlantic Ocean in the latitude 35 North.

There is no bottom to be discovered in the Ocean at any great distance from land.[28] For example, in leaving the West Indies Islands, no sounding can be found 10 or 12 miles from these Islands, nor till within about 100 miles from the Lands End of England, and the ship strikes bottom in about 150 fathoms; the depth of what are called soundings is what makes a cables length about one hundred twenty fathoms.

[27] Walker's hand lens was not of high enough magnification for him to discover the microorganisms (chiefly dinoflagellates) responsible for the phenomenon of bioluminescence. When he refers to "marine insects" in the next sentence, he means small marine animals, not insects in the present restricted sense.

[28] Walker means simply that the bottom of the deepest parts had not been found. Inability of early navigators to find the bottom off the east coast of the West Indies was due to the presence of the Great Trench.

Along the western coast of Ireland there are no soundings; nor from the coast of Guinea till within 30 leagues of the American land. Soundings are found at different places, at different distances from the shore according or low. In general it is a rule well known to sailors, that the water is always deeper on a bold and rocky coast than on a flat one.

We shall next attend to the

HEAT *of the* OCEAN

The only good experiments we have on the subject in different climates were made by the late Captn. Dalrymple, and published in one of the vols. of the Philoph. Trans. Here we find a series of experiments which he made in his voyage to India from the latitude of 23 North to 32 South. Every day at Noon he had a Bucket of water taken from the surface of the sea, in which he immersed his Thermometer; and from these experiments it appears that the heat of the Ocean under the line is generally about 80° Fahrenheit, and never above 84°. And from the same experiments this is farther deducible that in June, when the Sun is in the Northern signs, the water of the sea in the latitude 12° South is about 83° Fahrenheit.

Between the line and 12° North, the Thermometer in the water at the surface [stood] at 81°, 82°, 83°, and the highest at 84°; but under the line, it was only from 76° to 80°. Dr. Forster in many places under the Equator found the temperature equal to 74°. Captn. Ellis likewise made an excellent experiment upon the heat of the Ocean. In one of his Voyages in the latitude of 25° North he found water at the surface to be 84° in Fahrenheits thermometer in his bucket sea-gage, the heat gradually diminished as the depth increased, till it arrived at the great depth of 3900 feet. The Thermometer then stood at 53°. And though he let it down to the depth I formerly mentioned, 5346 feet, it stood regularly all the way at 53°, and it remained fix'd at the point, though it was let down above 1000 feet farther. It appears then that beyond the depth of 3900 feet, at least in that latitude, the heat continues the same as deep as we know, and probably all the way to the bottom. In consequence of his, we infer that the Ocean in the warm Climates diffuses cold, and in the cold climates it diffuses heat. High land is known everywhere to propagate cold; and of this we have a striking instance in the voyage of Lord Anson.[29] When he ranged the coast of South America, opposite the Andes, the atmosphere was mild, and even cool. But as he proceeded north to the latitude of 10° or 12°, he found it extremely hot and sultry. The reason of which is that when he was under the line he was in the immediate

[29] Lord Anson (1697–1762), Admiral of the Fleet.

neighbourhood of the Cordeliers, which rendered the atmosphere, even at that distance, cool when compared to what it was 10° or 12° from the line, but beyond the reach of the cold formed by these mountains, almost covered with Snow.

The author gives another reason, to wit, that on the coast of Peru, the violence of the sun beams is blunted by the clouds that perpetually intercept the rays of the sun. It is a dispute among the naturalists whether the ocean is ever congealed; or whether all floating ice is not drifted from land. Captn. Cook found a piece of floating ice in 50° South, whose height above the water was 200 [feet], and hence must have been 1800 feet thick. It has been thought that these icy islands were formed at land and have acquired increase of bulk from the falling snow and the dashing of the sea upon them; but since no land has been discovered by the Navigators between 57° and 67° south, we have reason to conclude this ice the production of the sea itself, except in the polar regions where the ice is of lamellated structure and contains, in cavities formed by the lamellae, a considerable quantity of very brackish water. On this account, the ice must be washed before it is melted down for using in food. As this is not the case in the polar regions, the difference probably arises from a greater or lesser degree of cold. The field ice is often very extensive and is discovered by our Greenlanders by what they call the Sea blink, or luminous appearance, at a great distance. The surf . . . runs so high as to prevent landing on coasts, running 15 to 30 feet high, and is so intermixed with air that no boat can live in it. It is so peculiarly conspicuous between the tropics and undoubtedly depends upon the trade winds giving a tone of current to the ice. The *Feicus natans* Linn., Sargasa, in some of the Southern seas, floats upon the surface of the sea for hundreds of leagues, and is probably the most copious vegetable of the globe. There are fishes [only?] known out of soundings in the temperate and frigid Zones.

The Submarine regions are only to be known by diving. Mr. Spalding of Edinburgh went down to the Royal Gorge and performed many feats of diving. He was a good mechanick, a tolerable Chymist and a courageous man. He fell a victim at last to his intrepedity, and, had he lived, would undoubtedly have enlarged our ideas of the submarine regions. He went in his bell as deep as 12 fathoms, and was consequently pressed by a weight equal to two atmospheres. At this depth the light was perfectly good, so that he could distinguish the very nails of the wreck he was down at. He went down in very high gales of wind, altho' the waves run very high at the surface, but he found everything still and quite tranquil at the depth of three fathoms.[30] The light was almost as good at 12 fathoms as upon the ships deck which he left.

[30] This is a very early recognition of the shallowness of the wave base.

He found that one gallon of Air English just served him for breathing a minute and no longer; altho in so compressed a state.

WATER SPOUTS

The phenomena that frequently occur at sea, but they have scarcely been so circumstantially described as to afford sufficient data for laying open the true cause.

These appearances are more frequently in the Mediterranean and tropical latitudes than anywhere else. They are frequently mentioned by our late circumnavigators. They are always preceeded by a dead calm. In the places where the spout is to arise, the sailors observe a large white spot; from this a column of water arises and is met by another apparently descending from the clouds. This column when formed is smaller in the middle than at top or bottom; and it rises in a Spiral manner from the sea. This column likewise moves along the surface of the sea and hence arises its oblique directions. When the water spout disappears, an impetuous and sudden wind springs up, blowing from several points of the compass at the same time. Then a column of water descends at once from the atmosphere into the sea in the form of a cylinder. The only probable cause of these water spouts is that wherever thick heavy clouds are wrought upon by contrary winds, they are compressed and made to descend in a stream seemingly occasioned by whirlwinds.

There is another sort of water spout, which is called the

TYPHON

This sort prevails only in the Sea of China. It is a huge column of water which is thrown up by the sea into the atmosphere, and is supposed to be owing to an explosion of some volcanic or submarine eruption, as is evinced by the constant sulphureous smell of the adjacent air; and there is frequently pumice thrown up along with it.

In many parts of the sea there are remarkable

WHIRLPOOLS

They have always engaged the attention of Voyagers. These don't appear to differ much, either in their kind or cause, from those which occur in quick running rivers. They appear to proceed from an irregularity at the bottom, from steep abrupt mountains or rocks, or from the confluence of two opposite currents.

The most remarkable whirlpool in the ocean is the Malstroom of Norway in the latitude of 68°; and which is said to have a diameter of about 20 leagues. These whirlpools are in the form of an inverted cone; and the water moves downwards in a Spiral direction. When anything gets within the Sphere of this vortex, it is drawn downwards in a spiral manner to the center; from thence to the apex of the cone, where it is then left at liberty to emerge.

It has been supposed, that whirlpools were occasioned by a great hole[31] through which the water descended to the interior parts of the earth; but as there is no evidence of such aperture, we must refer them to one of these causes:

1st. to the conflux of opposite strong tides; or

2nd. to the opposition made to the currents by huge rocks.

There is a very remarkable one in our own country at the coast of Colybrachan in the Western islands. It is situated in a sound between the islands of Jura and Skarba, which lies at the northeast end of Jura, and the Sound is remarkable for its turbulence. This sound is about two miles broad, and the whirlpool on the Skarba side is not far from Shore. Soon after the flood enters the sound, the sea is in great disorder and makes a hideous noise. This commotion increases till near the fourth hour of the flood, when the noise is loudest, and the sea is then most impetuous, and throws up every thing moveable. The waves rage with great fury and extend for several miles. Soon after the fourth hour of flood these violent motions abate, and about an hour before it ceases to flow, the sea becomes smooth, and the place of the whirlpool is not known. Soon after the return of the Ebb, the sea again becomes agitated, and its motions increase till the fourth hour of Ebb, when they are most violent and then they cease until the tide begins to flow.

The inhabitants say, that the attraction of the whirlpool extends to a great distance, but for an hour at high and low water, vessels may not sail near or even over it in security. They tell of a tall ship, coming within the sphere of its vortex, [which] was abandoned by its crew, and was at once swept and immediately cast up with the shells and sand from the bottom. There are several other stories told of its effects, but I have them not authenticated. The cause of the whirlpool seems to be a perpendicular submarine rock which opposes itself to the course of the current as it passes through the strait. The opposition it gives may cause these gyrations in the water, which we have

---

[31] A popular theory to explain the origin of rivers and lakes was that proposed by Kircher in 1678. He held that holes existed in the bottom of the sea and communicated with subterranean passages through which water passed and emerged high in the mountains, whence it flowed to form lakes and rivers.

described. It accounts for its being most violent during the highest tides; and it likewise accounts for the smoothing of the surface when the tide has ceased to flow, or when nearly flood; but it accounts not so well for the same smoothness at the ebb. There is another cause however that may account for this last, to wit, the diminished force and stagnant state of the tide at that time.

We come now to consider the subject of the

TIDES

which are commonly reckoned of five kinds, vizt.

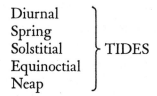

Diurnal
Spring
Solstitial ⎬ TIDES
Equinoctial
Neap

The Diurnal Tides

which are the common flux and reflux of the ocean, occuring twice in 24 hours.

The Spring Tides

are those which occur at the full and change of the moon.

Neap Tides

or the smaller tides, occurring at the moon's quarters.

The Solstitial Tides

Being the weak tides, they take their name from the considerable variation observable in the tides at the summer and winter solstices. And the last are

The Equinoctial Tides

or those which occur at the Equinoxes, and rise the highest and fall the lowest of any tides throughout the year.

OF *the* CAUSE *of* TIDES

I may begin here by taking notice of the superstitious causes that have been assigned in order to account for this grand phenomenon of the waters of the ocean. The Ancients knew nothing of the cause of this phenomenon. It was an old opinion, that the Alteration of the tides was occasioned by the influx of rivers into the sea; but this cause is altogether inadequate to the effect.

There was a theory proposed and started by Aconius[32] in the middle of the last Century, which was received by several Philosophers of the Perepatetic School, that the vicissitudes of the tides were owing to a motion of the earth from North to South, in the manner of the pendulum of a clock, twice in 24 hours; but it never was proved, and probably never will, that the earth has any such motion.

Aristotle, upon studying the cause of these, gave up the search in despair, and, in vexation, threw himself into the sea at Euripus.

There was an idea which was maintained by the famous Galileo, and many succeeding Philosophers, "That the tides were occasioned by the diurnal motion of the earth." But, if this were the cause, it behoved to follow, that in 24 hours, there should be only one flood and one ebb instead of two. The famous Descartes was the first Philosopher who took notice of the moon's having any concern in the tides. He maintained that wherever the moon came to the meridian it occasioned a depression of the waters of the sea, and consequently they flowed on the shores; and, that on here verging towards the horizon, they recoiled and ebbed.

About the end of last Century, a very celebrated mathematician, Dr. Hales, published a very elaborate mathematical treatise on the tides, and even thought it possible to account for them from the attraction between the moon and the waters. His theory hinged on what he understood to be the common center of gravity of the earth and the moon. He lived to relinquish this theory, but he was the first Philosopher who mentioned the important observation that there was an evident observation and attraction of the waters of the ocean by the moon wherever she became vertical. This opinion was embraced by Sir Isaac Newton and makes the basis of the Newtonian theory of the tides, to wit, "that whenever the moon comes to the meridian, she attracts and deviates and occasions the flood, and as she verges towards the horizon, the waters recoil and resume their place and occasion the ebb.["] This account, however, was incomplete; for there must have been high water in her opposite places at once. This Mr. M. Laurin[33] proved to be owing to the solar attraction. Vid. Buff: Translat. Vol. 1. Art. XII.

There is indeed one Phenomenon in the tides, which does not coincide with the Newtonian theory; a Phenomenon, which indeed, the incomparable author of that theory was not informed of, but with which the world has

---

[32] Perhaps a certain Acontius (1492?–1565), born in Italy, lived in England.

[33] Colin Maclaurin (1698–1746), professor of mathematics, University of Edinburgh, 1725. His *Treatise on Fluxions* (1742) contains an essay on tides.

since been made acquainted by many navigators; and that is that when the moon is in our meridian, and occasions the flood, as is supposed by attraction, the flood subsists in the same minute in the opposite side of the globe. That is, that when it is high water here, it is high water exactly at the same time with our Antipodes.

There is an observation which, at first view, does not seem to coincide with the Newtonian theory, though I am persuaded that if it was fairly examined it would be found perfectly correspondent with it.

After this account of the causes that have been assigned to account for the Phenomena of the tides, we now proceed to consider the Phenomena themselves.

It is now a general Opinion that there is a current set round the globe, particularly in the intratropical parts, and for some space beyond them, found east to west. It has been noticed that the Current on the north side of the Equator, while the Sun is in Cancer, has a direction to the north, and when the sun is beyond the Equator and gets into the Southern signs, this current is two or three degrees to the south of west, and we have great reason to suppose that the same thing happens in the southern hemisphere.

We find that wherever there are open straits anywhere on the globe that are open to the east and west, such as the straits of Magellan and Anian, these the sea passes with a considerable current from east to west and without returning.

There seems in different climates to be a remarkable resemblance between the equinoctial and spring tides; for the tides of the equinoctial regions generally rise to six feet, and on extraordinary spring tides, they swell six inches higher. But in our climates, where the tides are high, for example in the river at Bristol, they are elevated usually to 60 feet of perpendicular height. Now [this] is just ten times the height of the tides under the Equator. Then at Bristol, the spring tide rises to 65 feet in height, so that there is an increase of just ten times what the increase is under the Equator. And it is probable that similar proportion obtains all over the globe; the spring tides being 1/2 more than the ordinary floods, providing other interfering circumstances do not cause a change.

Whenever the tide of the Ocean is uninfluenced by the land, the winds or any other cause, it rises regularly for 6 hours. At the expiration of this period, it continues stationary for 12 minutes, and then the ebb takes place and continues for another six hours, after which there is again 12 minutes, when no alteration takes place. After this the flood again begins. Of this each days tide is 48 minutes later than that of the preceding; and as the moon is just so much later in the 24 hours in getting to the meridian of any given place, on account of its moving slower than the Earth, this affords a beautiful and satisfactory proof of the Newtonian doctrine and constitutes one of the most remarkable Phenomena of the tides. We know that the Earth moves faster than the moon by 13 degrees in 24 hours. We know also that the moon requires one hour to move 15 degrees. Therefore it must happen that the moon in 24 hours will be left behind the Earth 48 minutes, and that whenever the moon is vertical at any particular place, and at any particular time of the day, the succeeding day she must be and actually is 48 minutes later in becoming vertical at that particular place. This forms a striking confirmation of the Newtonian theory of the tides, and indeed amounts almost to a demonstration.

This short interval of 12 minutes, at the time of flood and ebb, is commonly called Acquistorium, but in some parts it varies in its duration. In the Mediterranean, the Acquistorium is said to last an hour.

There is a Phenomenon concerning the tides, which is perfectly ascertained, and that is, that neither the flood nor the ebb move equal spaces in equal times. This has been very accurately observed in one case, and I have here a table containing the results of several accurate observations that were made with regard to it at Plymouth harbour, when the flood amounted to 16 feet of perpendicular height.

	Hour	FLOOD Feet	FLOOD Inches	EBB Feet	EBB Inches
During the first hour of flood the water rose	1	1	6	1	6
2nd hour it rose	2	2	6	2	6
3rd hour it rose	3	4	..	4	..
4th hour it rose	4	4	..	4	..
5th hour it rose	5	2	6	2	6
6th and last hour it rose	6	1	6	1	6

Total 16

And, what is very remarkable, it was found to fall at ebb precisely in the same proportion, as shown above.

From this table we see that the greatest velocity both of the flood and ebb is about the middle of these periods, for the water rises and falls most at half flood and half ebb. But the real velocity is extremely various in different places, depending greatly on the ly of the shores, though it is every where unequal during the ebb and flow. I don't know if it is any where so regular as in this example.

Where rivers run into the sea, the flood is short, and the ebb proportionally long, particularly about the mouths of large rivers, for example, along the shores of the river St. Lawrence. The tide ebbs 7 hours, and the flood continues only 5 hours. But in opposition to this, on the coast of Aquitaine in

France, the flood makes only 4 hours and the ebb continues for 8 hours. Such are the inequalities to be observed in many places with respect to the duration of the tides. But then this certain observation takes place that whatever variation there may be either in the time of the flood or ebb, both of them always make up 12 hours, and in general, whenever variation takes place the ebb takes up more time than the flood.

The highest spring tides, though they occur in consequence of the moons being vertical, yet they do not correspond to the fullness of the moon. It is commonly at the third flood after full moon that the tides are highest in general. The highest spring tides make the lowest ebb, though with some exceptions. Sometimes there will appear a most remarkable ebb, greater even than what happens at spring tides. This happens by a gale of wind from the land. It is well known to sailors, and they call it an outlet.

The highest annual spring tides don't happen on the day of the Equinoxes, but at the full or change of the moon immediately following. There is another very remarkable observation with respect to the tides, and an observation that requires more attention, tho it has been noticed so frequently that there is little reason to doubt of its truth, and that is, that from the month of March to the month of September the evening [tides] are considerably higher than those of the morning. While on the contrary, from the month of Septem. to the month of March, the morning tides are higher than the evening tides. Of this Captain Cook made an useful application in one of his voyages along the coast of New Holland. He had the misfortune to run his ship upon a coral rock on which she stuck fast in spite of every exertion to disengage it. But from excellent talent of observation with which he was endowed, he had observed that the evening tides in the adjacent seas were three feet higher than the morning tides, and taking advantage of this circumstance he made every preparation for another trial at midnight which proved successful. By this means he saved his own life and the lives of his valuable crew.

The highest tides that are known in the northern hemisphere occur between the latitudes of 45 and 55°. The same has been lately discovered in the southern hemisphere, tho' there the tides are less than in the northern. When we get so far north as 60°, the tides diminish. The daily tide [is] imperceptible in the Atlantic till within ten leagues of the European land. It is a remarkable circumstance that in the latitudes where the tides are highest . . . the range of the Barometer is greatest.

In the Island of St. Helena, the tides don't rise above the height of one foot. At Barbadoes and the adjacent islands, the common tides are four feet, five inches in height. At Bermudas they are five feet. At Capt Corse-Castle on the Coast of Guinea they arise to six feet. In Plymouth harbour the tide amounts to sixteen feet, but the highest tides that are known in the world are on the coast of France. At St. Maloes, and in the river at Bristol, (as I formerly mentioned) the common spring tides amount to sixty, and the Equinoctial to 65 feet. But when the Atlantic is urged on by a strong westerly wind, the tide sometimes amounts to seventy feet. At St. Maloes in France the water rises 84 feet, which is the highest rise of the tide of which we have any accounts. The tide at Liverpool is very high, tho considerably less than at Bristol. It rises there 48 feet at the equinoctial tides. But so far as I know, there is no tide to the north of Liverpool so high; the tides on the east and west coasts of Scotland amount from 20 to 28 feet at the Equinoxes. At Leith the Spring tides rise to 15 or 16 feet, or more if the water is forced up the firth by a high N.E. wind.

They gradually diminish in height as you proceed towards the north and the highest in Ireland is only to 16 feet.

We shall next take a general view of the

Progress *of the* Tides
in the
British Seas

The general flood sets from the west. It comes from the Atlantic on the shores of Britain and Ireland. At the southwest end of Ireland it is high water when the moon is in the meridian at 3 O'Clock in the morning. The tide then comes on, so that at 5 O'Clock in the morning it is high water at the land's end of England. At Plymouth it is high water at 6 O'Clock. At the Isle of Wight it is high water at nine O'Clock. At Dover at half past 10 O'Clock.

Off Dunkirk it is high water at 12 O'Clock, and in the Texel it is not high water till 3 O'Clock afternoon. This great flood which enters the British Channel from the west, after passing the Straits of Dover, rushes into the German Ocean and then proceeds northwards. We shall now trace the progress of the flood at the other extremity when it passes the north west extremity of Ireland, it divides into two branches. One branch enters the Deucaledonian Sea and falls in among the Hebrides at the same time it passes by the Irish coast and through the Mull of Kintyre with prodigious rapidity. After passing this strait, it goes South through the Irish Channel till it reaches the Isle of Man, and then encounters the tide coming up St. Georges Channel. This is well known to the Sailors in the Irish seas, the confluence of these tides forming a surge or current called the Race of Aep, which is exceedingly perilous at certain times.

The other branch passes through the Pentland firth and goes among the Orkney and Shetland Islands, and when it enters the German Sea, it proceeds southwards. It meets the other great branches at the Dogger bank. That bank which stretches across the German sea from Hamburgh head to the Dutch coast. It is very probable that that large Bank has been formed by the confluence of these tides. This current seems to be a consequence of that great Current[34] which is put in motion by the trade wind and is urged against the coast of Brazil and Terra Firma. It is thence directed to the Gulph of Mexico, whence it is urged by the Cape Florida along the coasts of the United States, as far north as the Banks of Newfoundland, whence it is thrown over towards Ireland. The West India beans and seeds that are found on the west coast of Scotland afford ample demonstration.

The currents in the eastern seas are entirely regulated by the prevailing monsoon, always keeping its direction.

We come now to make some observations on the

TIDES *of the* INLAND SEAS

Such of these seas as possess any greater degree of saltness than the Ocean don't derive it from their inland situation, but owe it to their neighbourhood, to extensive strata of fossil salt. The tides in the inland seas are small, though they are remarked. For example, the Caspian Sea, that great body of inland water which extends about 900 miles in length and 400 in breadth, has no outlet, tho' it receives the waters of many great rivers and of rains, and this is explained by Doctor Pallas to be compensated by the greatest of evaporation from its surface.

In this great body of water there is not the least appearance of tide. It is scarcely perceptible in the Euxine Sea;[35] this sea is emptied into the Propontis by the Thracian Bosphorus. The current in this great strait is equal in velocity to a rapid river, and it is remarkable that there is both an upper and an under one in contrary directions, the upper running into the Propontis. In the Mediterranean there is everywhere some degree of tide, but its greatest height does not exceed two feet.

This is not easily reconciled with the Accounts we have of the extensive Lakes in North America. We are told that in Lake Huron, which has [no] communication with the sea; at least no such communication as can in

[34] The Gulf Stream; an early recognition of its influence.

[35] The Black Sea.

degree affect the tides, there is a perceptible flood and ebb; that in Lake Superior there is a flood and ebb of about 6 feet;[36] and that is as regularly produced by the moon as the tides of the Ocean.

And yet as far as we can find, there is no such thing observable in the Caspian Sea, which is a larger body of water. In the Baltic there is a tide of about one foot, which indeed seems to depend entirely on the attraction of the moon and not on its communication with the German Ocean. In winter or in autumn at least, when the Baltic is filled by the rainy season thaws, there is a constant current sets out of it into the German Ocean, which is not influenced by the tide. In the heat of Summer the Baltic is known to shrink under the level of the German Sea, from which, at that time, there is a regular current, which sets into the Baltic. But in autumn, when the tides of the ocean cannot affect the Baltic, because of the current setting from it into the German Ocean, then the tides are regular, which shews evidently that they must depend on the moon.[37]

The Black Sea is 900 miles in length, and 200 in breadth, and its waters are almost fresh, or are only a little brackish. We are not well acquainted with the state of the tides in it.

In the Thracian Bosphorus, the tide rises about 10 inches. The current, running from the Euxine into the Propontis, is said to double the water on the surface running from the Black Sea and that of a few fathoms deep running from the Propontis.

The Palus Maeotis has very little of tides and communicates with the Euxine by the Cimmerian Bosphorus, and its waters are nearly fresh.

The Dead Sea of Judea is 70 miles long and 16 miles broad, and among other small rivers, receives the water of the Jordan. It has no sensible outlet; its waters are very salt and abound with bitumen, which, hardening upon the shores, forms the Asphaltum. It contains in it no fishes. Some Philosophers have supposed that the waters of Jordan were conveyed away to the Red Sea by some subterraneous passage. But there is no reason for this conjecture, for an accurate traveller has computed that the water poured into the Dead Sea from the Jordan is about 6,090,000 Tons; and it may be calculated that this is the evaporation from its surface.[38]

The level of the Mediterranean is higher than that of the Atlantic. It

---

[36] See n. 22.

[37] The Institute of Marine Research, Helsinki, Finland, reports that in the most favorable positions there is a tide of about one-half foot. As a rule there is a constant surface current moving outward that is independent of the tide and that is compensated by a bottom current moving inward. (Personal communication.)

[38] This is a very early attempt to calculate the rate of evaporation from a body of water.

appears to be still doubtful whether the Mediterranean gives more water to the Atlantic than it receives from it. In the Gut of Gibraltar there are 5 remarkable streams. The middle of the Strait is occupied by the current which runs from the Atlantic into the Mediterranean (and which is not affected by the tides of the Atlantic). On each side of this there is another current setting in the same direction, which is always affected by the tides. Between each of the last and the shores, there is another current setting from the Mediterranean into the Atlantic, and these two last currents occupy by far the greatest part of the Strait. The middle current, however, is the strongest, for when the others cannot be perceived, it is still evident.

We come now to another branch of our subject, the history of

SPRINGS

There are different sorts of springs known by different names.[39]

The Subterranean Springs are such as circulate in the earth without making any eruption at the surface, but are met with every where when we dig to a sufficient depth.[40] These afford all our pit and pump water and are of very different qualities at different places. In deep mines I have always found them equally pure with the finest spring water.

Those are termed Day Springs which bubble or boil up at the surface. In long extended plains there are few or none of these day springs, as there the water proceeds not from a height sufficient to force its way through the earth, for water will rise as high as its fountain head,[41] but no higher. This is called the Libration of water, and from this comes the use of water pipes by means of which the water may be carried over considerable heights provided these are lower than its fountain head.

Those are termed Temporary springs which flow the greater part of the year but are dry in the heats of summer. These run near the surface and are soon and greatly affected by a rainy or dry season, and they are sometimes

[39] The section on springs is in general of very high quality. It shows a splendid and modern understanding of the source of spring water, the cause of ebb and flow of some springs, the significance of hydrostatic head, the temperature of spring water, the relation of the temperature of spring water to the temperature of the atmosphere, the movement of underground water in relation to the structure of the strata and caverns, the relation of underground water to the "hanger" wall of mines, and the mineral content of spring water. Walker was clearly an expert on springs. His first published paper was on the composition of a well producing mineral water.

[40] This refers to the intersection of the groundwater table.

[41] The hydrostatic head. When Walker speaks of the "libration" of water he is referring to that state wherein the subsurface water is in hydrostatic equilibrium.

observed to depend on the rise or fall of Rivers. Thus when a river is high you will find a number of springs in the neighbourhood, but when very low these disappear.

There are other springs termed Periodical. These subsist in all the Alpine country as in Switzerland. They break forth regularly in the month of May and dry up in the month of September, and evidently derive their water from the melting of the Snow in summer.

There are others termed Intermittent Springs, which have a kind of rise and fall, ebb and flow in a regular manner, and in a pretty determined time.

There is one of these at Forbay, that ebbs and flows 5 or 6 inches, and about 16 times in an hour.

But the most remarkable spring of this kind is in the Diocese of B——— in West Friesland. It is called ——— or the Boisterous spring. It loses itself twice in 24 hours, it returns once in each six hours with great noise and violence.

We know not any thing particular with respect to the cause of these intermittent springs, but they are very probably owing to the alternate compression and expansion of air in their subterranean channels.

As to the quantity of water poured by springs travellers have described to us the famous spring at B——— [Vaucluse?] near Avignon. It moves three mills within half a mile of its source. Travellers however have not mentioned the quantity it pours in a given time but from the above account we may easily suppose it very great. We are informed too that even in this copious spring the water is well known to increase by rain and to diminish by dry weather.

But the most copious spring we are acquainted [with] is that remarkable one in Flintshire and called St. Winifred well. It turns 23 mills within a quarter of a mile from its source, and fortunately the quantity of it has been exactly measured, and it is found that it pours no less than 90 tons in a minute.

The origin of Springs is a physical question which has been much agitated among Naturalists.

Many suppositions have been assigned to account for them, but none altogether satisfactory.

Aristotle considered them as being occasioned by vapours condensed in caverns of the earth. Springs however are everywhere found, but few such caverns are known to exist.

The famous Descartes thought that springs owed their origin to the evaporation of sea water by means of subterranean fire; and our Countryman Lord Stair, who was a great admirer of Cartesian philosophy, is in his

Physiologia Nova at great pains to establish this opinion. Becher[42] the celebrated German Chymist thought they were occasioned by the subsidence of the ocean into the earth, and from thence ascending in the form of vapour.

But in accounting for springs in this way the authors have always been greatly incommoded by an objection which occurs to every one, vizt. If the water of springs are thus derived from sea water either by evaporation or subsidence, what is to become of the prodigious quantity of sea salt, of which there is no vestige to be found.

The celebrated Dr. Halley formed another opinion concerning the origin of springs. He went for an astronomical purpose to St. Helena. On the summit of this Island in the night time his glasses and papers were so much wet by the dew that it was with great difficulty he could accomplish his purpose, but it led him to think that springs were derived from Dew, and this opinion he published long ago in the Philosophical Transactions.

Dr. Woodward in his theory of the Earth, endeavours to establish the opinion that all springs were derived from his supposed subterraneous abyss of water in the center of the globe.

The ingenious Mr. Derham supposed that they derived their origin from the [circulation] of sea water in the bowels of the earth; but there are many instances to show that this cannot be the case. Thus in the Bermudes, and even in many of the sandy shores of Britain, there are many springs elevated but little above the level of the sea whose waters are perfectly sweet whereas the water in the ground a little nearer the sea is salt.

In order to form any idea of the cause of springs we must take a view of their phenomena, and of the established facts we possess in their history.

First as to their heat. It has alway hitherto been the general opinion, tho' indeed a most erroneous one, that the heat of springs is the same in all countries, and indeed it is but of late years that this opinion has been invalidated, and it is still held by many eminent people. It appears however from many experiments and observations lately made that the heat of springs varies considerably in different parts of the globe, according to the climate and the general degree of heat in any particular country.

Farther it is now found that the heat of springs is always the medium heat of the country where they exist, that is of the Atmosphere.[43] The

---

[42] Johann J. Becher, *Physica subterranea.*

[43] Various explanations had been given to explain the origin of hot springs as well as the varying temperature of cold springs. In this discussion Walker accurately and clearly interprets the temperature of spring water. It must be remembered that at this time many people believed that spring water came from the center of the earth and that rain water could not penetrate

medium heat of the Atmosphere in any particular country is best ascertained, by observing the heat of midday and of midnight for a whole year, but if the medium heat is to be determined by a single observation in the day, 8 O'Clock in the morning is the most proper hour because in general it gives a result of the medium heat of 24 hours better than any other. Accordingly by a register accurately kept of the Thermometer for near 30 years in the Customhouse here at 8 O'Clock in the morning. It appears that the medium heat about Edinburgh in all that time was exactly 45° and a fraction of Fahrenheits thermometer, and accordingly 45° or 46° is the medium heat of all the springs in the neighbourhood of Edinburgh.

In the South of England the medium heat of the Atmosphere is 2 or 3 degrees more, and in consequence of this the heat of springs there is 48°. The heat of springs in Pensylvania was likewise fix'd at 48° by Dr. Franklin. We know exceedingly well the medium heat of the climate about Paris both from the state of the thermometer above ground, and the subterranean heat in the cave of the observatory. In the cave of the observatory the heat is stationary all the year through at 53° of Fahrenheit; this likewise is the medium heat of the Atmosphere and of all the springs.

There is but one other place in which this has been observed, vizt. at Upsal by the celebrated Bergman. He informs us that all the springs in that neighbourhood have the heat of 43°. He does not indeed inform us of the medium heat of the Atmosphere but we may conclude it is pretty nearly the same.

In short the heat of the springs in any country generally gives the medium heat of the Atmosphere and the medium of the stationary subterranean heat.

I have always regretted that the heat of springs has no where been observed within the tropics, and it is but very lately that any observations have been made on the subject, but within these two years we have an account of several very accurate ones in the Island of Jamaica, where the heat of a great number of springs was ascertained, and these were found to correspond exactly with the medium heat of the climate.

So long ago as the year 1761 I had occasion to notice that in this country the heat of the springs was gradually changed according to their height above sea level.[44] Thus in the high mountain of Hartfell in Anandale, springs at the foot of it stood at 45° of temperature. Higher up on the

---

downward more than a few inches. Walker relates the temperature of springs to that of the median heat of the atmosphere, using data collected from various parts of the world including information from Pennsylvania obtained by Benjamin Franklin. Walker's recognition of this condition is one of the very earliest notices in geological literature.

[44] This is a very early recognition of the influence of elevation on the temperature of springs. As Walker points out, he made the observations in 1761.

mountain there were others which were at 44°, 43°, 42° and one which was near the summit at 39° while that at the foot of the mountain was at 45°. This is now found to be the case likewise in Jamaica where however the difference in proportion to the level is much more considerable. On the sea shores of that Island the springs are very hot; they become gradually cooler as you ascend from the sea, and on the mountains their heat is very moderate.

It is pretty certain that all the springs in this country are warm in the time of frost, the reason of which is very obvious. Spring water is endowed with the subterranean degree of heat, but in the winter time all the surface water is cooler. There are very few springs however which are indeed mix'd with surface water, but in the time of the frost when all the surface water is bound up then the springs acquire a higher temperature.

It appears that springs are capable of acquiring heat by running through calcareous strata.[45] The first time I noticed this was at Bath. There the river Avon runs through a calcareous country and is so heavily loaded with chalky matter as almost always to have a whitish colour. In a very intense frost I found the water of this river considerably warmer than that of any other of the neighbouring rivers. The same thing has since been established especially by the experiments of Bishop Watson. In the chalky country about Cambridge he found the heat of the springs always 50°, whereas in general the medium heat of the climate of England is only 48°. There are in Cambridge shire and in other chalk countries very deep wells sunk into the chalk. In these wells Bp. Watson generally found the heat 50°, 51° and even 52° when at no very great distance and out of the chalky country the heat was only 48°.

The same seems to be the Cause of that remarkable degree of heat in some of the springs of Derbyshire which all run thro' calcareous earth. At Matlock there is one the heat of which is 68° and in that neighbourhood there are many other springs which though commonly reckoned cool are from 50° to 54°.

Besides, calcareous earth spring water is capable of deriving heat from some other Causes. Thus we are informed by ——— of Derbyshire of several springs there that derive their heat from passing thro' strata of pyritical matter and it would appear that all the hot springs of France do run thro' strata of fossils in which is pyritical matter.

The circulation of water in the bowels of the earth is facilitated by the various inclinations of the Rocks and strata and their numerous fractures,

[45] This observation needs further confirmation. What Walker says is probably true, but the cause is not readily apparent.

and we generally find the subterranean springs flowing in the direction of the strata.[46]

The subterranean waters,[47] in this circulation underground, observe generally the direction of the strata, and of the innumerable fissures and fractures that run through in all directions. The subterranean waters found in mines are unquestionably the sources of all our springs, whence they proceed along all the various strata of the mountains and issue forth at various outlets; and subterraneous waters are always found most copiously near the surface of the earth. For at Leadhills, at the depth of 116 fathoms, the miners find the quantity of water diminishes as they go downward 600 to 700 feet from the surface. Whence, a presumption arises that all the springs and rivers proceed from the superior atmosphere, and by no means from a central abyss, which, we have reason to think, does not exist; for in all low countries, as in fenny countries of England; in low Countries of Holland, there are no regular perennial springs. . . .

There is seldom any spring found where there are deep earthy strata, or where there is a great depth of loam, sand, clay, marle or fullers earth. In such tracts there are always few or no springs. Some of these substances, such as clay, marle and fuller's earth, resist the ascent of the water, sand absorbs it, and both serve to prevent the water from issuing in Springs at the Surface. On the other hand, in rocky countries Springs are very numerous. . . .

It has generally been assumed tho' without foundation that Rain water never goes to any considerable depth in the earth. This error is owing to the want of exact observation. I have known a fall of Rain in 24 hours penetrate 270 feet of Slate, sand, granite and clay; and in the mines of Nithsdale it is well known to the miners that a fall of rain will sometimes penetrate 50 fathoms. Accordingly in Mines all the subterraneous water proceeds from the roof of the mine, as when they are working a vein it comes from the hanger[48] as it is called.

It is likewise no less remarkable that all mines tho' never so much distrest by water, are always dryest in the time of frost because then all the water near the surface is locked up. Accordingly where there are water engines for working mines in the time of frost they are laid idle, but then they have fortunately little to do.

It is likewise well known that all mines and levels are destructive of

---

[46] The influence of structure on subsurface circulation of groundwater is recognized here.

[47] This and the following paragraph are from D.C. 2-18.

[48] The roof side of the vein, the side hanging above the head.

springs, which serves to shew that they are all derived from the surface.[49] There is likewise another erroneous opinion on this subject, vizt. that perennial springs always pour the same quantity of water. But upon enquiry it will be found that there is no perennial spring [how]ever pure or steady that is not affected in some degree by a long continued tract either of wet or dry weather. In this country the quantity which perennial springs pour is most diminish'd in the months of August and September, because at these times, the earth in consequence of the summer heats is in its dryest state, and this was well known to the inhabitants of Edinburgh when scrimp of water.

It appears evident that those springs which cease entirely to flow in dry weather do entirely depend on rain for their support.

Springs situated in plains are generally the largest, and it will always be found on every mountain that the higher you go up the springs always diminish in the quantity of water which they pour. Accordingly there are very few springs seen in all low and small Islands, but in those Islands however small if there be highland, springs are copiously found. This is the case with St. Helena and with our remote Island of St. Kilda in both which the land is high and rocky and accordingly they have plenty of springs.

It has been commonly reported that there are springs on the tops of Mountains, and this was used as an argument that they were not derived from the Atmosphere. But though I have been at a great deal of pains in searching of such yet I never found any.

We are told there is a spring on the summit of the Bass. But in reality it is about the middle of it and above that there is a prodigious rock.

In like manner it has been supposed that there is a spring on the top of Dumbarton Castle rock. There is indeed a small spring on that rock but it is near the middle of it. There is no spring on the top of Edin[burgh] Castle, but a copious one at the foot of it.

We are likewise told of collections of water or small lakes on the summit of mountains, but in this country I know of none unless one in the south of Scotland on the top of Queensbury hill, and this is only a sort of bog on

[49] When mining operations intersect the zone of saturation the groundwater table is usually lowered in the immediate vicinity. With the lowering of the groundwater table springs may disappear. This is a very early recognition of the principle involved. Walker derived much of his information from personal field observations. Studies of mines added to his knowledge of both groundwater and structure. A common opinion about springs was that they owed their origin to the "upward exhalation" of deep subterranean waters. It was supposed that some springs issued from mountain peaks. Walker went to the field to study the temperature and flow of springs and to see whether he could find one on the top of a mountain. He effectively demonstrated the falsity of the concept of upward exhalation, and properly related springs to subsurface water.

the summit situated in a small hollow and is evidently occasioned by the fogs and rains, the mossy ground preventing the water from sinking.

In like manner it has been supposed that there is a spring on the top of table mountain at the Cape of Good Hope. But by the account Sparman gives of it, who went on purpose to see it, it is probably similar to that on Queensbury mountain.

Hot Springs

In the Island of Iceland is the Geyser, the most remarkable of any of these hot springs in the world.[50] Dr. Lind[51] told me that he had measured it by a Quadrant, and found it throws no less than 92 feet high into the Air, an effect which is only produced by the explosive force of confined steam and is almost from 188 to 212° of Fahrenheits Thermometer. Others like it are found there also. It contains dissolved in [it] a true siliceous stalactites which it deposits in its course, and is found to be a silica nearer to Chalcedony than to Quartz or any other siliceous stones, thus siliceous stalactites, the only one known, is formed in these springs. The Earth deposited by these Springs in the form of a powder is a curious matter, for it is really a pulverized siliceous substance. The water of the Geyser is very remarkable for it is the only water or known spring that we know of which is impregnated with Genuine siliceous Earth.

I met with this description in an ancient Iceland Chronicle of great antiquity, which relates and records the history of facts from the 9th. century upwards. It mentions that a great earthquake having happened and that to a dreadful degree; and that then a large spring of boiling water took place, which was never known before, on this occasion; and that it rose a certain number of Danish ells in height; and that this was in the 11th. Century, and makes the first appearance of the famous Geyser.

From these Phenomena and facts well established we are led to draw the following conclusion with respect to the origin of springs, that the waters of the atmosphere, hail, rain, snow, fogs and dew, are more than sufficient to supply all the subterranean water and springs, and indeed the floods of any large river occasioned by rains is many times the quantity of all the springs in the neighbouring country.

We come now to mention some vulgar and indeed they may be termed

[50] This and the next paragraph are inserted from D.C. 2-24. Here Walker recognizes the existence of geysers and geyserite. This is probably the first recognition of the spring-deposited material called geyserite. This is usually credited to J. C. Delametherie in 1812.

[51] James Lind, M.D. (1736–1812), College of Physicians, Edinburgh; made investigations in the Hebrides; constructed portable wind gauge.

Philosophical errors on this subject. Such as that the water of springs proceeds from a central fire, that they rise from the sea, that they issue from the summits of Mountains, that sea water ascends to form springs, that Rain water cannot descend above 10 feet in the earth, that perennial springs always pour the same quantity of water, that the largest springs are always at the greatest depths in the earth.

We come now to the natural history of

RIVERS

Upon the European continent the Danube is by far the largest river. It is computed to run 1500 miles in a straight line. In Africa the River Niger runs in a straight line 2400 miles. The River Nile in a straight course runs 2520 Italian miles, and with its curvatures it may be estimated at full 3000 miles.

These rivers however great are nothing to those on the continent of America, compared to which (says Thompson) our floods are rills.

The River Amazons runs no less than 1350 Spanish leagues in a straight line. This is unquestionably the largest river in the globe, and next to it is the Mississippi.

It is commonly supposed by naturalists that all rivers have in the course of ages cut out their own channels. I think however there are many instances in this country to shew that this cannot be the case.[52]

You will find several instances in Scotland and some in England.

There is a remarkable instance at Bristol. There the river Avon below the town of Bristol enters on St. Vincents rocks, the rocks on each side of it being exceedingly steep and about 300 feet in height. It is not to be supposed that the river had made here its own channel, or that it would attack this rock when it might have spread itself upon the whole south of England.

It appears then upon the whole that the channels of the Rivers have not been fortuitously formed, and especially in all rocky countries, they have most probably been at least coeval with the rivers themselves. There is a remarkable incident in some rivers [of] their occulations. Instances of this are pretty frequently to be met with in the Highlands of Scotland in the case of small rivulets.

There is likewise the case with the River Rhone, which below Geneva falls under ground, runs for a considerable way in the subterraneous channel and

[52] The origin of valleys was not one of Walker's strong points. Although he understood a great deal about the work done by running water, he did not recognize the importance of either downcutting or lateral erosion by streams. His interpretations may have been influenced by the fact that he lived in a glaciated territory where valley forms had been greatly modified by glacial action and he could not conceive them as the result of stream erosion.

then again emerges. There are many instances of this likewise in the eastern parts of Europe, and it is to be attributed where there is a cavernous structure of the earth.[53]

We shall now make some remarks on the Inundations of Rivers and the first and most interesting of these is the overflooding of the Nile in Egypt.

The rise of the Nile is never perceptible till after the summer solstice. It rises indeed for some days before but a few inches in the 24 hours, but at the height it rises considerably and frequently from 15 to 48 inches in the 24 hours.

Its greatest height as observed and ascertained by Dr. Shaw in the course of 29 years took place always between the 15 of July and the 19th of Sept. and the rise from first to last occupied about 100 days.

On this subject we have a very curious account by Herodotus so that if we could be assured of its authenticity would help us to judge the formation of the lower Egypt.

He assures us that under the reign of Meris king of Egypt which however was 900 years before his own time, that the Nile, when it rose 8 cubits[54] in height was sufficient to inundate the whole of the lower Egypt; these 8 Egyptian cubits amounting to 13 feet 8 inches French measure. He also informs us that in [?] time when in Egypt the rise of the river to 15 cubits or 25 feet 7 inches French measure was sufficient to inundate the delta or lower Egypt.

The Emperor Julian in one of his letters written in the year 362 informs us that on the 20th of September of that year the river only rose to 15 cubits.

In the present age a rise of 18 cubits is reckoned a standard of abundance because it is necessary to innundate the whole of the lower Egypt, and accordingly when it only rises to 16 cubits it is very insufficiently watered, and then its produce is so considerably diminished that according to the law of the Turkish government, Egypt does not on that year pay the annual tribute. But at present the Nile sometimes rises to the height of 22 cubits and then it is reckoned rather detrimental by remaining too long on the ground and thus preventing the sowing of crops.

The height to which the Nile rises is always marked on a pillar called formerly M——— now known by the name of the Nilometer. Before the rising of the river it stands 5 cubits in the water and at the least during the height of the inundation at 16 cubits more or 27 feet 4 inches French measure.

It appears then on the whole that all of the lower Egypt has in the course of a number of ages had its soil and surface greatly heightened.

[53] The relationship of "lost rivers" to underground cavernous systems is clearly portrayed and understood.

[54] In English measure 18.22 inches; 17.4 inches in Roman cubit.

To account for this remarkable phenomenon of the Nile's inundation many different causes have been assigned. The Ancients attributed it to the Elesian winds which blowing in the spring and summer season in the opposite direction of the Nile's course; they supposed that they dammed up its waters, and when they went off occasioned its inundation.

Strabo supposed it owing to the summers rains, others have ascribed it to the melting of the Snow in the hyperborean regions. A late writer in France Mr. B——— has a peculiar opinion concerning it. He had performed some experiments on the slow running rivers of France erroneously concluding that the Nile had only equal velocity, and therefore he supposed that its inundation arose from the melting of the Snow in the former summer at or near its origin. The rivers in France which he observed run only four leagues a day. This must be a creeping course indeed, but the Nile is a river of great rapidity, and thus as in all other rivers is much increased by a flood.

The Ancient opinion of the wise men of Egypt concerning the overflowing of the Nile was that it was owing to the clouds at the polar regions being attracted by the mountains of the moon and hence being discharged on the land of Africa in the form of rain.

But from our knowledge of the situation of the river we have now no difficulty in accounting for the true cause of this appearance. The most distant heads of the Nile lie in the equatorial regions and its inundations definitely depend on the heavy solsticial rains that fall in the equatorial country of Abyssinia.

The waters of the Nile in a flood must have a [great] deal of slimy matter but this will be made rather fine owing to its long course.[55] This sediment [is] deposited in the delta or lower Egypt where the waters are at liberty to spread out, and this is evidently the cause both of the great depth and fertility of the soil there. The Ancients were so well acquainted with this that they stiled the lower Egypt "τὸ δολον τοῦ Νίλου."[56] But if we allow the report of Herodotus to be true then we are fully entitled to form the following conclusion—that the Delta has in the course of 3284 years increased 14 cubits in height, and indeed if we are inclined to allow its formation from the sediment of the Nile this is a very moderate degree of quickness.

In like manner we know that land is formed at the embouchures of all rivers flowing in a flat country. Aristotle observed that the Palus Meotis was not so deep in his day as it had been formerly, and in all probability it is not

---

[55] A clear concept of the reduction of large fragments to small during the process of transportation. The source of river sediment and its ultimate deposition at the mouth of the river to form deltas is splendidly illustrated.

[56] Or δῶρον τοῦ Νείλου—the Gift of the Nile. See Herodotus, Bk. II, chap. v.

as deep now as it was in his time, which is owing to the great quantity of sediment brought into it by the great river Tanais.

Ostia in Italy, which was formerly the Port of Rome and which was close to the sea shore, is now in consequence of the sediment deposited no less than 3 Italian miles removed from the sea.

The Po likewise has formed no less than five separate islands at its embouchure in the Adriatic. There are other rivers known to inundate in a similar manner to the Nile, and for the same causes; all those which have their sources near the line as the Congo, Senegal, etc.

But there are others which inundate for a different cause as the Volga which annually rises between the 25th of April and the 20th of July. At Astracan [Astrakhan] this river is in its ordinary state 2200 feet broad and of a very considerable depth, but on the 27th of May in the year 1770 Professor Pallas observed it to rise no less than 7 feet 9 inches, and this rise is evidently owing to the melting of the hyperborean snows.

The quantity of sediment in the waters of any river vary according to the nature of the soil through which it flows, and to the degree of flood.[57]

I do not find that the quantity of sediment contained in the water of a river has been ascertained by anyone but by Planeus, he took the water of Rimini in a flood, and from the evaporation of 51 inches of this water was afforded 3 inches of dry sediment, here then the earth was to the water in the proportion of about 1 to 17.

On the other hand those rivers that flow from Lakes or Inland seas have their water generally clear and with very little sediment. I remember long ago that all the people in this country before the conquest of Canada were greatly alarmed on account of the sand banks which they supposed were at the mouth of the river St. Lawrence. But it was judged by a certain gentleman, a priori that this could not be the case and his reasoning was very sound and plausible; for said he the river of St. Lawrence proceeds from a chain of Lakes in which it will deposit its sediment, and will pass thro' the last of them pure, consequently no banks will be formed at its embouchure; and this was actually found to be the case.[58]

Those rivers, which are fed from springs, are always warmer in winter than those obtaining their waters from rains and melting snows. But those

[57] I should like to see this concept called "Walker's Law." In a few words Walker summarizes the basic facts about the load of a stream. The statement has been paraphrased many times by writers unaware of its source.

[58] This deduction by Walker is true and represents a very early recognition of the role of lake basins as sites of deposition. The incoming streams laden with sediment drop their load, whereas the outlet streams are clean and more or less free of sediment.

rivers, which run over beds of calcareous earth, are warmest of any. This I have often observed in the rivers of England.

Water, therefore, taken from a river in a high flood, must always be muddy, containing sometimes no less than a seventeenth of the whole earthy matter. We must not be surprised then that great changes are produced at the afflux of large rivers.

We come now to make some observations on

CATARACTS

Several of these are described by travellers on the continent but not with precision; as the Cascade of Tourny in Italy which is said to fall 300 feet. The fall of ——— in Switzerland is said to be 900 feet high, and ——— in Switzerland 1100. But when we hear of these we are not to think them proper Cataracts but only steep falls of water from rock to rock over precipitous mountains, and of these we have several to as great and even greater heights in the highlands of Scotland.

But of a real Cataract the greatest we are acquainted with is that of Niaga[r]a in North America; this was formerly described by a French writer Father P——— [Hennepin?].[59] His accounts however of every thing are wonderfully exag[g]erated, so that he is now called by his countryman, *Le Grand Menteur*. He affirmed it was 600 French feet high, and it was in consequence of this report that the minds of men were struck with such awe.

But we have had of late more accurate accounts of it by a French officer who makes it 137 English feet and still more lately by Capt. Carver[60] who measured it by a cord and makes it 140 English feet. The rock over which this Cataract falls is a plain, simple gray limestone of which there is a specimen in this room.[61]

We come now to make some observations on

LAKES

Of these the Lake of Geneva is the largest in the European continent. In length it is 18 leagues and at its greatest breadth 5 leagues. It rises remarkably in its level from the end of February to the middle of June by the melt-

[59] Louis Hennepin (1640–1701), French missionary and explorer in North America.

[60] Jonathan Carver (1732–80), *Travels throughout the Interior parts of North America*.

[61] This is the oldest reference I know to the kind of cap rock forming Niagara Falls. It is remarkable that Walker had been able to obtain a specimen of this "simple gray limestone" and have it on display for his students. Perhaps one of his American students or Benjamin Franklin sent it to him.

ing of the Alpine snows, in general this rise is from 12 to 15 feet. In many places it is likewise very deep, in one place no less than 500 fathoms, at least, it having been sounded to that depth without finding bottom. Mr. Saussure gives us lately a very curious fact with regard to this lake, but in a loose manner; he says that throughout the whole year, at the depth of 120 feet, the heat of the water is always stationary. But he has omitted to tell us of the precise degree of this stationary heat.

The Lake of Geneva like all other Alpine Lakes is the deepest in proportion to its size.

The Lake of S——— in Switzerland lately measured by Count Razonousky was found to be 960 fathoms deep.

The generality of Lakes receive rivers or rivulets and generally emit some, as in this country Loch Lomond, Loch Tay, and Loch Ness, and such likewise are many of the North American Lakes. Others receive Rivers but do not emit any as the Caspian and Dead Seas. There are some likewise which receive none but send one out as the Pu [———] Lake in Tuscany and several in North America. This happens with those lakes where the spring and rain waters are superior to its evaporation.[62] Some others neither receive nor emit any river, we have a Lake of this kind in this count[r]y vizt. ——— Lake; this is likewise the case with Thracima Lake in Tuscany, and it happens when the spring and rain waters are equal to its evaporation.

The American Lakes are remarkably large, as Lake Champli [Champlain] which is 80 miles in length, and ——— in breadth. The Lake of Bourbon is likewise 80 miles long. The Lake of [Huron?] 200 in length and 100 in breadth, Lake [Ontario?] 280 long and 40 broad, ——— 600 long, Erie 300 by 40 miles, and Lake Superior no less than 1600 miles in circumference. There is one Island[63] in this lake no less than 100 long and 40 in breadth, and yet the waters of all these lakes are sweet and wholesome.

We have an account from the French of a Lake[64] in North America West from Hudsons Bay the water of which gained 3 feet in height during the space of 7 years, and in the course of the ensuing 7 years subsided to its former level; but whether this is constant or not and what is its Cause we are totally ignorant.

There are some Lakes which do not freeze in the winter time as Loch Tay in Perthshire, which in the time of frost constantly smokes, and no snow lies

[62] Walker recognizes that lakes may derive part of their water from subsurface sources such as springs. He also understands that evaporation may equal supply, so that no outward-flowing streams exist.

[63] Probably Isle Royale, which is about 44 by 7 miles.

[64] Lake Winnipeg.

within six feet of the verge of the water, the river Tay runs out of it unfrozen and runs five miles without freezing. There is a little river called the Lyon which joins the river Tay, and the Ice of which is dissolved in that river in the course of a mile or two. Loch Tay is of a very great depth, no less than 130 feet in some places, and so likewise is Loch Ness which likewise does not freeze, and so we may conclude that those that do not freeze are of a very great depth. But there is one Lake, Loch Lomond, which is one part shallow and in the other deep; this lake covers about 17800 acres and is situated only about 22 feet above the level of the tide, from about the middle downwards, from Lus to the exit of the water of Leven it is about 20 fathoms deep, but above Lus it is all of a great depth, especially immediately under Ben Lomond within a gunshot of the shore where it was measured by Dr. Cullen and found to be 98 fathoms. Here it does not freeze, but all the shallower part is very frequently frozen.

There have been several Causes assigned for this Phenomenon of Lakes not freezing. Some have attributed it to their great depth, but the Lake of Geneva which is much deeper than any of them is known to freeze almost every winter.

Ca——— and B——— seem to have been well acquainted with this Phenomenon as occurring in the Lakes of this country, and ascribed it to bituminous matter with which the water was impregnated; but we know this is not the case.

Dr. Plot thought that these Lakes did not freeze by means of some insensible streams of Salt, or some internal fermentation. These however are but mere words and serve rather to make the matter more obscure than to throw any light upon the Cause.

But the great depth is the cause of this, at the Lake of Geneva, where the thermometer has been plunged into the waters, the mercury subsides, till the thermometer arrives at 150 feet deep, and then it falls to its utmost extent, and the mercury never moves. Therefore, the waters beyond that depth remain of the same temperature however far you may exceed the depth. The same is to be presumed in respect to the Lakes in Scotland. So that all the water in the inferior parts is of the same temperature, and this prevents the superficial waters from freezing; for no sooner it cools than it must subside and give place to warmer waters below.[65]

The true Cause seems to be combined of the great depth of these Lakes and that their water almost entirely consists of springs.

[65] Walker compares Loch Tay with Lake Geneva in regard to freezing. This paragraph, inserted from D.C. 2-24, is a clear statement about the density stratification of lake water by temperature turnover. It is probably the earliest statement on density stratification in literature.

There are some of the Lakes of Asia which contain a very great proportion of Saline matter, and form a saline crust very evidently sea salt, these Lakes by sinking in the summer time, sometimes leave the salt half a Russian ell in thickness.

Others contain a great quantity of Glauber salt. I shew you some native glauber salt brought from a lake in the neighbourhood of Astracan taken up by Professor Pallas and sent to me.

It is likewise highly probable that in some of these Lakes the fossil alkali exists in great quantities in an uncombined state.

There still remains a particular species of Lake which is of a very extraordinary nature; of these there are several in different parts of the continent, but it does not appear that they have ever been exactly described or their Cause ascertained. These Lakes are remarkable in this, that their water appears and at a determined season totally disappears.

The most common of these is in the country of Kamnivo called the Serchnick Lake. It is about 2 German miles in length and about one in breadth; it is from 2 up to 60 feet deep and yet the whole water in the month of June subsides thro' apertures in the earth and totally disappears, the waters return in the month of October by the same apertures and on their return bubble up with some violence often to the height of 6 or 7 feet. It is also well stored with fish and these both go and return with the water. The first account we have of this Lake was given by Dr. Brown[66] who travelled into that country and whose travels were published in the year 1683. After that a particular treatise was written on it in 1689 by Baron Bazoni,[67] but all that he says with regard to the cause of this Phenomenon is that it is owing to subterraneous syphons.

It is pretty clear that the water of this Lake must retreat to some other Lake or River above ground by some subterraneous channel, and probably is brought to such a level by a wet or dry season.[68]

The particular place to which the waters go, is as yet unknown, but might be easily estimated by observing the fish, or by throwing into the Lake before it disappeared a great quantity of any light substance as wood, cork, etc.

There is another Lake of this kind called the Lake of C——— in Dalmatia,

[66] Dr. Edward Brown (1644–1708), president of the College of Physicians, London, *Brief Account of Some Travels in Hungaria* . . . (1673). Cerknica is southwest of Ljubljana in Yugoslavia. Kamnivo is no doubt Kamnik, a village near Ljubljana.

[67] Unidentified.

[68] That the level of subsurface water fluctuates with rainfall is a deduction of major proportions. Walker seems quite aware of the importance of underground caverns in limestone terrain and of both the lateral and the vertical movement of subsurface water.

3 Italian miles in length. Its waters disappear about the month of June and subside thro' apertures of the earth, one of these apertures is no less than 20 feet in diameter; they again return in the Month of October and fill the Lake in a few days; a prodigious quantity of fish rises out of the earth with them, and returns again at the retreat of the Lake.

In the summer season both these Lakes are perfectly dry and sown by the inhabitants with grain which is in general reaped before their return in October.

It would seem that the [rocks] of the earth in the Eastern parts of Europe and especially between the Danube and Adriatic were very cavernous.

There is one observation which I find made by Aristotle and also by Pliny. There is a particular species of fish enters the river ——— at a certain season of the year, and without being observed to return, it is seen at another time in the Mediterranean. Both Pliny and Aristotle suppose that this fish which is called the Sardina goes by subterraneous caverns into the Adriatic.

The Island of Kerson on the coast of Dalmatia contains the Lake Jeseron of the same kind in general; it is 14 feet deep, but it abandons the Lake pretty regularly every 4th year. It is sown the first year by the inhabitants with Indian corn the 2d and 3d with wheat, but they sow nothing in it the 4th year.

There is in Russia a Lake of this kind called the Lake of E——— which is regularly 3 years a lake and 3 years dry ground and is then cultivated by the inhabitants.

This extraordinary appearance of refluent Lakes is as yet but very imperfectly described, and no cause is assigned.

The fabric of this Globe in every part is remarkably fitted for the constant circulation and equal distribution of water thro' the whole; the waters of the ocean are elevated to the earth, from the earth they are raised into the Atmosphere, and from thence again fall upon the earth, which every where rises gradually in its level from the sea, so that water runs from the most inland parts of the earth to the ocean. Every part is thus supply'd with water, and if there are any little inequalities they serve only to shew the great wisdom [with which] the whole is conducted.[69] There we may behold the footsteps of that divine power which every where pervades the works of nature.

[69] Here the hydrological cycle in its complete form is recognized.

We come now to another branch of our subject vizt.

GEOLOGY

OR

The Natural History of the Earth

Here we shall first consider the general fabric of the Globe.[1] Most of the Ancients and many of the Moderns have held that this globe has originated in water and subsided from a watery fluid. The celebrated Descartes was the first who gave to it a fiery origin. He ascribed its origin to a star or dry globe of fire gradually deprived of its light and heat.

The Count de Buffon's theory[2] is expressed but in different words and in a worse manner. Descartes however seems to have spoken with appearance of candour on this subject.

The first thing observable in general with regard to the Fabric of the globe is its disposition into Land and water. It would appear in general that a certain proportion of Land and water is necessary for supporting the economy

[1] In D.C. 2-18 the first paragraph of this section reads, "In this branch of our course, the first thing remarkable that presents itself is the Fabric of the Globe itself. And hence we may observe its admirable dispositon into land and water, and the proportion that exists betwixt these to answer the economy of nature: the gradual rise of land from the sea, so necessary for the proper circumstance of the globe. We commonly guess at the interior parts of the earth. The deepest mines are but scratches on its surface, when we consider the globe itself."

[2] Georges Leclerc Buffon (1707–88), *Histoire naturelle.*

of the habitable world; accordingly we find the water to the earth nearly in the proportion of 3 to 1, and this great watery space among other uses appears to us necessary for supporting the sufficient proportion of vapour.

The great body of Land we now pretty certainly know lies towards the north; till of late years it was supposed that there was a great Southern continent and this was supposed necessary for preserving the equilibrium of the earth; but we now know that no such continent exists but that in the high Latitudes of the southern hemisphere there are nothing but small Islands.[3]

Every where too it is found that the earth rises gradually from the sea shore to the more inland parts, which is necessary for the regular and constant circulation of water.

There have been many opinions and indeed very wild ones concerning the central parts of the earth. Kircher[4] supposed that all the central parts of the earth were occupied by great and separate masses of fire, and that it was by means of these that the water of the oceans was evaporated, rose through the earth in the form of steam, and thus gave rise to springs, and that in some places these masses of fire had certain spiracula and these were the volcanoes.

Another ingenious man Hernis[5] the Swede gives us a different account. He supposes all the central part to be occupied by boiling water, that the water proceeds from it so as to occasion springs; he supposes likewise that between this central abyss and the surface, there exists great hydropylacia from which all the waters on the surface of the earth and in the Atmosphere are derived.

It was presumed by both these Philosophers that a central fire was absolutely necessary to account for the subterranean heat,[6] but if this were the case then at the greatest depth the heat should be sensibly greater, but we know that this is not the case for when we go to a certain depth in the earth the heat is perfectly stationary and evidently depends on the medium heat of the Atmosphere; hence we conclude that all the subterranean heat is derived from the Atmosphere and from the Suns rays.

Hernis's idea of a central abyss gave rise to Dr. Woodward's[7] theory of

[3] As most others of the day, Walker did not suspect the existence of Antarctica.

[4] Athanasius Kircher (1601–80).

[5] Urban Hjerne or Hjarne (1641–1724), *Acta et tentamina chymica in regio laboratorio Stockholmiensi* . . . (1712). Identified by W. Odelberg, librarian of the Royal Swedish Academy of Science, Stockholm.

[6] Walker uses the word "subterranean" to refer to anything below the surface of the earth, not necessarily the central core.

[7] John Woodward, *An Essay toward a Natural History of the Earth.*

the earth, in which he endeavours to account for many of the phenomena by means of this central abyss.

Van Helmont[8] supposed the central parts of the earth to be occupied by what he calls a *sabulum aquorum.*

Becher[9] the German Chymist likewise adopts the idea of great hydropylacia.

We know that as yet man has not penetrated one mile into the earth. The deepest mines are still inferior to this. There is one in Joachimstad in Bohemia 2100 feet, and we only know of one deeper than this, in the Tyrolese, the exact depth of which I do not know.

Mr. De Luc[10] went with the Barometer to the depth of 1309 feet, and I have gone to the depth of 892 feet at Lead Hills which is the greatest depth of the mines there, and there all is solid rock. We never meet with caverns of any importance nor earthy strata.

In short the internal and central parts of the earth are quite unknown to us.

Dr. Pallas[11] and other modern philosophers suppose that the chief ingredient in the central parts of the earth is granite, and this is exceedingly probable if you join with granite all the other primary or coeval rocks.

Sir Isaac Newton offered an opinion from theoretical reasoning concerning the specific gravity of the earth, he thought upon the whole that this was as 5 or 6 to one when compared to water.

This conjecture was put some years ago very meritoriously to the test of experiment by the Royal Society, and accordingly Dr. Masklyne[12] took up his residence on Chehalion in order by accurate experiments with the Pendulum, to determine how far it was attracted by the Mountain. In consequence of these laborious experiments Drs. Masklyne and Hutton[13] make the specific gravity of the earth as $4\frac{1}{2}$ to 1, compar'd to water. They likewise suppose, for it does not appear they had made any experiments on the subject, that the specific gravity of the rock of which Chehalion is composed as compared to water [is] $2\frac{1}{2}$ to one.

[8] Jean Baptiste Van Helmont (1577–1644), student of heat and gases.

[9] Johann Joachim Becher, *Physica subterranea.*

[10] Jean Andre De Luc (1727–1817), *Relation de différents voyages dans les Alpes du Faucigny* (Maestricht, 1776).

[11] Peter Simon Pallas, *Observations sur la formation des montagnes et les changements arrivés au globe, particulièrement de l'Empire Russe* (St. Petersburg, 1777).

[12] Nevil Maskelyne (1732–1811), *Phil. Trans. Roy. Soc. Lond. Ser. A* 55 (1785): 495. Maskelyne determined the specific gravity of the earth by experiments on Schehallion, a 3,547-foot mountain near Perth, Scotland.

[13] Charles Hutton (1737–1823), *Phil. Trans. Roy. Soc. London. Ser. A* 48 (1778): 331.

But I am inclined to think this is erroneous, and that they had imagined it of the same specific gravity with the Secondary rocks in the south of England as Portland stone. But we know very well that the rocks of the highland mountains and especially that of which Chehalion is composed, are much heavier.

The specific gravity of the earth therefore, is probably somewhat different from that which it is made by Dr. Masklyne for by his computation the specific gravity of the globe to that of the rock of Chehalion is as 9 to 5; but we know that the schorl rocks of which Chehalion is composed, compared to water are not less than 3 to 1.

But this computation being admitted vizt. the earth is to the rock of Chehalion as 9 to 5, then indeed a very curious and interesting question arises, of what can the internal parts of the earth be composed? We know very well that on the earths surface there is no fossil whatever whose specific gravity is to water as 6 to one but what partakes of metallic nature, every other kind in specific gravity at or above 6 to 1 owes its gravity to metallic matter; and on the other hand those fossils which do not contain Metal are generally in specific gravity as $2\frac{1}{2}$; 3; 4, and the *Spatum ponderosum*[14] itself between 4 and 5 to 1, and it is not absolutely certain that this last is free from metallic matter.

Dr. Hutton solves this question by supposing the internal parts of the earth to be of a metallic nature. But this is a mere supposition and indeed rather an improbable one.

It is indeed likely that the internal parts are more dense than the external merely by compression, and indeed we find in the strata of any considerable rock those parts at the bottom more compact and heavy than in the upper part of the rocks.

But I am rather inclined to think that the specific gravity of the earth is not quite so much as determined by their computations, and so we will have not this to account for.

Concerning the general fabric of the globe, we may here take notice of its

WESTERLY ELEVATION

In all parts of the world the Mountains are high and precipitous towards the Western side and they decline and shelve away towards the East.

To illustrate this we may begin at hand. In the Kings park here there is

---

[14] Barite (sp. gr. 4.5). Walker recognizes that only the metals have a high specific gravity and that rocks become more dense and compact with depth, but he cannot quite bring himself to believe that the high specific gravity of the earth means the equivalent of a metallic core.

evidently a Westerly elevation, the hills gradually shelving away towards the East. The same likewise is the case with all the Islands in the Firth of Forth, and in some degree with all the mountains of Scotland. I do not pretend to say that this elevation and shelving away of mountains is exactly West and East, but always towards these quarters.

It is in consequence of this Westerly elevation that the water shed[15] thro' the whole of Scotland rises from the West; which may be seen by the course of the rivers, at least all the large ones, as the river Tay which runs more than 70 miles, and has its remote feeders, even within 4 miles of the Western ocean.

There is but one considerable river the Clyde which is an exception to this. It rises near the Western ocean; it attempts to run to the East, but cannot, and is therefore turned into the Western Sea, but the general level of the country is contrary to this, and at one place, £100[16] would make the Clyde run to Berwick.

The same observation holds in England. The water shed there lies near the western side of the Island.

In short all Britain is evidently an inclined plane from West to East, and accordingly the mountainous tract in the West side of the Island has its valleys all narrow and contracted, while in the East they are large, extensive and exposed.

This observation of the Westerly elevation may be carried into the continent of Europe in general. Thus the mountains of Norway to the North Cape all shelve away to the East. The same is the case in Switzerland, all the Alps being steep towards the West, and shelving away towards the East; and it is likewise the case with all the particular mountains as Mount Buet.

The course of the Danube shews the general inclination of the country from West to East.

The same is the case in Asia, there the most elevated sources of the rivers as Caucasus, run to the east, likewise the most considerable rivers as Euphrates, Indus, etc. Even all China is well known to incline from West to East.

The observation likewise equally applies on the continent of America. In North America there is not a single river running West, 1100 miles west from Niagara. The central partition or water shed is there far to the West; in Latitude 45° it lies very near the Pacific ocean; there we find the sources

---

[15] This is an early use of the term watershed to designate a drainage divide. It is used again when referring to the drainage in North America. He also uses the term feeders in referring to tributary streams feeding a master stream.

[16] I.e., labor to the value of £100?

of St. Lawrence and of the Mississippi. Nor in the whole of Mexico is there any large rivers. They are all very short in their course, compared with those which fall into the Eastern sea as the Amazons and the river Plate.

The same thing likewise takes place in the very extremities of America as in Terra del Fuego.

This fact concerning Westerly elevation has pass'd unnoticed by the common Theorists of the earth, but is certainly a leading circumstance in the Natural history of the globe.

Here it would be needless to enquire into the Cause of this Westerly elevation, as it would require to be established by more extensive observation.

The next thing to be taken notice of in Geology[17] is the Natural History of

MOUNTAINS

The celebrated Mr. Ray and Dr. Hooke were the first who ascribed the origin of Mountains to earthquakes.[18] They have been since followed by many others.

A great number of late and present philosophers have attributed them to Volcanoes.

But their regular form, their regular chains, their general direction, their continued ridges, the salient and receding angles, their strata and the matter of which they consist, evidently shew that they never have arisen from such partial or contiguous Causes as Earthquakes or Volcanoes, but that they owe their origin to a Cause more universal and much more uniform. It is upon the whole much more probable that they have been formed in water than either by earthquakes or volcanoes.

The Ist thing to be noticed with regard to mountains is their Shape. Notwithstanding the great diversity of figures observable in Mountains, I think it possible to divide all of them into seven different sorts.

The Ist form is the Conical mountain or Sugar Loaf; as North Berwick

[17] In this section Walker classifies mountains into seven different types based on form, a very early attempt at classification of geomorphic features. He notes the relationship between the long axis of mountain chains and the lie of continents and islands. In addition he calls attention to the classification of mountains as primitive and secondary, the first composed of primitive strata and the latter of secondary strata. These two kinds of mountains, he says, "have been formed at two very different æras, the one certainly previous to the other." The secondary mountains and strata are "superincumbent" upon the primitive strata. This appears to be the oldest use of "æra" as a time term in geology.

[18] John Ray (1627–1705), *The Wisdom of God Manifested in the Works of the Creation* (8th ed.; London, 1722). Robert Hooke (1635–1703), "Discourses on Earthquakes," written in 1688 and published in the *Opera posthuma Robert Hooke* (London, 1705).

Law in this country, and Loganhouse Hill the highest of the Pentlands, and the great Mountain in Teneriffe.

The IId form is the Peaked Mountain. These are broad at the base, grow narrow toward the top, but at the very summit are hunched or peaked; such is Arthurs Seat, Ben Lomond in Dumbartonshire and many other of the highland mountains.

The IIId form is the Rounded mountain or Round back'd mountain, for example that range of hills in the southern part of Mid Lothian called the Lammermuir hills, each of which is rounded, while the Pentland hills opposite to them are generally either conical or peaked.

The IVth kind is the Ridgy mountain or such whose summit consists of one lengthened out extended ridge, of these there are several in the West of England where it is called the Cwm[19] as in Lancashire, and of this kind there is a remarkable one among the Pentland hills called House hill.

The Vth form is the Table mountain. This is nothing else but a truncated Cone, and there are several of that appearance in this country as that remarkable hill in Anondale called Burnswark and a ridge in the Isle of Sky consisting of five mountains of a very considerable height which pass with the Sailors by the name of McLeod's tables, as they are on McLeod's estate; and such also is the Table mountain at the Cape of Good Hope.

The VIth form is the faced mountain, where it shows at the very summit an abrupt or nearly perpendicular face, as Salisbury Craggs in this neighbourhood; and in the highlands most frequently those where basaltes occurs.

The VIIth form is the Acuminated Mountain. Such are the great group of hills in the northern part of Scotland in the county of Ross where they all terminate in spires. This figure of a mountain generally takes place where it consists of Schistus or slate.

These seven species of form in Mountains are exceedingly distinct; so that every hill on the Globe may be reduced to one or other of these shapes.

It is a remarkable circumstance in the case of all mountains disposed in chains, that the chain always runs in the direction of the continent, promontory or island in which the mountains are. This is the case in Britain. There is a chain runs along the Western side of Britain North and South.

There is likewise a chain on the shore of Norway from North to South, and one which runs through the island of Iceland. The vast chain of the Uralian mountains forming the natural barrier between Europe and Asia

[19] Cwm was used by Walker to refer to an extended ridge, probably a misuse of the native term. Cwm (in English Coombe) is a Welsh word for the head of a valley surrounded by steep slopes, as a cirque.

runs exactly South and North from the source of the river [Ural] to the frozen seas.

Nova Zemla is only a South and North excursion of these mountains.

The Peninsula of Kamachatka runs south and North from 62° Lat. to 51° and its chain of mountains runs through the middle of it in the same direction.

The Andes too have their ridges lengthened out in the same direction with the continent.

The Mountains in the Southern parts of North America run from 30° to 40° Lat. parallel to the coast and about 300 miles from it; and in like manner the river Mississippi has a range of mountains all along its course.

In short it would appear that the general direction of all the Mountains in the globe is South and North or nearly so.

The conical or peaked mountains are generally connected with ridges, while the roundback'd mountains are usually insulated, or detached from ridges.

There is an observation concerning mountains made by an ingenious gentleman in Switzerland Mr. Bouquer.[20] He had passed from Switzerland to Italy 10 or 12 different times by different roads thro' the Alps. It is an observation which I am confident is founded in nature and makes a great foundation to the System of Mr. Buffon, and that is, that when two chains of mountains run paralel, one of them throws out a salient angle which is always opposed to a receding angle in the opposite chain; in many parts of Scotland likewise this is exceedingly striking.

Mr. Buffon ingeniously draws a rational conclusion in consequence of this; that they have evidently been formed in water.

Notwithstanding the evidence now related and all that may be adduced in support of this fact, it is absolutely denied by some late writers as the Abbè Fortisse,[21] Mr. Dillon[22] etc., but it does not appear that any of these gentlemen have had sufficient opportunities of viewing mountains.

All contiguous mountains are generally of a similar shape. This I believe takes place all the world over. In our own country it is exceedingly observable. This is the case likewise with all bodies shaped in a fluid in separate masses. It is the case with dispers'd Clouds in the Atmosphere, with all the sands and banks formed by the sea, rivers or lakes.

[20] Possibly Louis Bourguet (1678–1742).

[21] Alberto Fortis (1741–1803), naturalist, *Mémoires pour servir à l'histoire naturelle et principalement à l'orycytographie de l'Italie et des pays adjacents* (Paris, 1802).

[22] John T. Dillon, *Travels through Spain with a View to Illustrate the Natural History and Physical Geography of that Kingdom* (London, 1780).

It is reasonable to think that in the progress of time mountains may be gradually worn down, because there is such a quantity of terrestrial matter thrown down from them by the rains; hence the highest mountains are generally the most rocky, and all mountains are most rocky towards their summits.

The gradual attrition of the tops of mountains [is] thought to be considerable by the descent of the waters. There is, no doubt, a great quantity of earth and terrestrial matter annually, carried to the valleys from the tops and sides of mountains, which are dislodged by waters; but this is not the only source from which the earth is derived. There is another, vizt. vegetation. The great number of vegetables that grow and rot yearly on the surface of these mountains, add to the surface everywhere and in course is carried to the valley below by the rains and winds.[23]

The degradation of Mountains is owing to different causes such as the falling of rocks, the undermining by the frost, and the wind, rain etc.

However in general it would appear that this degradation is chiefly confined to the mere earthy matter. In the high mountains of this country the summits are generally hard rocks covered with crustaceous Lichens. These are always exposed to the action of wind and rain, but do not lose perceptibly of their height, the loss being confined to the more terrestrial matter by the rains, and this terrestrial matter is again restored to them by the rains, the consequent vegetation and the decay'd plants returning to their former earth.

I come now to a distinction in Mountains, to which I would beg your particular attention, because of its great moment in the Science of Mineralogy, and in the natural history of the earth; it is that distinction by which mountains are divided into primitive and secondary. And accordingly the strata of which these consist have been termed primitive and secondary strata, and it has been acknowledged by all modern writers that these have been formed at two very different æras, the one certainly previous to the other.

They are likewise known by other names. The primitive mountains are called by some the primæval mountains. By the German Mineralogists and Naturalists they are term'd *gang* which is the German word for a vein, because in these the metallic veins are always found; and accordingly the French call them Montagnes aph———. Others call them vertical mountains

---

[23] This paragraph is from D.C. 2-18. In this section, which deals with the gradual degradation of mountains, Walker recognizes that the rate of erosion is slow. The reduction of mountaintops he assigns to the work of water, wind, temperature changes, gravity, and vegetation; these he believes, are capable in time of accomplishing the reduction of mountains. The words "attrition" and "degradation" indicate a modern approach to the processes of erosion and the destruction of mountains.

from the disposition of their strata; by some English writers they are called mountain rocks, as of these all our highest mountains consist.

The Secondary Mountains are termed by Ferber, Baron Born[24] and others superincumbent mountains, being generally found placed on the primitive strata. Their strata are likewise called horizontal strata, because they are generally placed in a horizontal direction.

But the most obvious distinction between primitive and secondary mountains arises from the different materials of which they are composed; these I shall afterwards have occasion to notice, at present I would only point out the materials by the most common English names by which they are known.

The primitive mountians or strata therefore are composed of the following fossils, Quartz, Feldspat, Jasper, soap rock, Lapis olaris,[25] Amianthus,[26] Asbestus, Slate, touch stone,[27] porphyry, serpentine, granite, whin rock, and Basaltes; all these rocks likewise are composed of Schorl or Mica.

The Secondary mountains or strata on the other hand are composed of a prodigious variety of fossils extremely different in their nature, as Lime stone, Shell marle, marble, portland stone, Alabaster, gypsum, Iron stone, sand stone, mill stone, and coal.

The primitive mountains are by far the highest; all the Alps in every part of the world consist of primitive rock. The primitive mountains or strata are never found disposed on an earthy or secondary stratum. But on the contrary the secondary mountains and strata are always found superincumbent to and spread out upon the primitive rocks.

The Strata of the primitive rocks are all vertical. That is they are generally placed at an angle to the horizontal of between 60° to near 90°.

On the other hand the secondary strata are very commonly found quite horizontal or inclined at the lesser angles of 25°; 35°; 45° and from that to 50° or 60 degrees but never above 60°.

In consequence of this vertical position of their strata, the primitive mountains are always the steepest, and have always less variety of rocks and less variety of fossils in an equal extent of country than the secondary mountains; thus in the highlands of Scotland you will travel 20, 30, or 40 miles without seeing any but primitive rocks, but in the low country where secondary strata abound you will meet with great variety of fossils in the space of half a mile.

[24] I. von Born and J. H. Ferber, *Briefe über mineralogische Gegenstände, auf seiner Reise durch das Temeswarer Bannat, Siebenbürgen, Ober- und Nieder-Hungarn* (Frankfurt and Leipzig, 1774).

[25] *Lapis ollaris,* an old name for soapstone (Wallerius, 1747).

[26] Amianthus includes white kinds of asbestos.

[27] A black variety of jasper.

In the primitive mountains the strata are never so thin or so numerous as in the secondary.

The primitive mountains are the general repository of all metallic veins. It is in mountains and strata of this kind that all the precious metals are found; to this rule indeed there are some exceptions, but these are very rare and the metals are never in equal quantity in the secondary rocks.

There is another remarkable and well established distinction between primary and secondary mountains, which is that the primitive contain no extraneous fossils. By extraneous fossils is meant petrified organiz'd bodies, or plants and animals in a petrified state. On the other hand these petrifactions are found in prodigious quantity all the world over in the secondary strata.

Another remarkable distinction between the two is that the primitive mountains and strata contain no sporadic matter of secondary strata.[28] The word sporadic is a term of art in Mineralogy. It is very concise, but useful as supplying the place of a great deal of circumlocution. All those fossils are termed sporadic which are included or imbeded in any stratum. Now in all or in the greatest part of the secondary strata, as of Marble, Limestone, Sandstone etc. we very frequently find masses of primitive rock included; but in the primitive mountains there are never any such masses of secondary rocks.

As I have already noticed the secondary strata are always spread out upon the primitive, but we never find the primitive superincumbent to those of a secondary nature.

In short the [composition?] and similarity in the ridges of mountains, their general direction from South to North, the similarity of shapes in the contiguous ones; the similarity of their materials and direction and inclination of their strata, and the regularity of their veins, all demonstrate that the Mountains of this globe have been formed by one general and uniform Cause, and that the primitive and secondary mountains have been formed at two very different æras.[29]

A third kind of Strata are called Accidental strata. These are the effects of the present and recent operations of nature. These are called by Mr. Kirwan when they form hills, tertiary hills.

[28] What Walker calls a "remarkable distinction" is in fact a remarkable observation. Here he recognizes that pebbles of primitive strata can be and are found in the secondary. He has also noted that extraneous fossils (organic bodies) are abundant in the secondary but absent in the primary. Only in recent times have we been able to find fossils in Precambrian strata.

[29] In this paragraph and the one preceding we note that Walker deduced the relative ages of rock strata by observing their relative positions. He has distinguished well between those rocks that we now refer to as Precambrian and post-Precambrian but recognizes only two times of mountain-building.

The accidental strata are generally earthy. Such are most of our strata of Clay, Loam, mould, sleech,[30] shell marle, sand, grit and gravel, in short they comprehend all those strata that are formed by the ocean or rivers, or by the showers of Mountains, they being very frequently found at the sides of mountains or at the foot. And to these likewise may be referred all those strata formed by recent or present volcanoes.[31]

Notwithstanding the degree of evidence which we have, you will find several writers and even some very late writers who endeavour entirely to abolish the distinction between primary and secondary mountains. But it is one which I am confident is founded in nature, and therefore must continue to stand its ground.

It would appear in general that the mountains of the globe are cœval with it, and it is fanciful to suppose that this globe ever wanted them, or that they have been formed in the course of succeeding ages.[32]

In short the general economy of nature absolutely required them to prevent the stagnation of water, to supply springs, to circulate freely the water from the more internal parts of the earth to the sea, to afford variety of soil and climate to the various plants and Animals, to produce the necessary current in the Atmosphere, and to support the general healthfulness of the earth.

The next thing to be considered is the

GLACIERS

In different parts of the world where there are mountains of a great height, their superior parts are frequently covered with perennial snow and Ice, these make a remarkable figure in Switzerland.

So far as I can find the lowest of these glaciers is 6000 feet above the level of the sea, there and between that and 7000 feet they usually begin.

Mr. Fox[33] informs us that in the valley of Chamouni he could touch the standing corn with one hand and the glaciers with the other.

The Glaciers are found of a great depth to 50, 60 or even 100 feet, which is only seen when they happen to be broken on steep and precipitous rocks.

Their surface is generally waved which very probably is originally owing

[30] A Scottish term referring to all types of mud. It is commonly applied to sea mud, river mud, and slime.

[31] Again Walker separates groups of sediments according to their time of origin. Alluvial fan material is recognized as "showers of Mountains."

[32] This and the following paragraph have been transposed from the end of the section on glaciers to maintain the continuity. Walker's remark here is far from true. Every geologist will recognize that no mountain system is coeval with the origin of the earth.

[33] Perhaps Charles James Fox (1749–1806), English political leader.

to the drifting of the Snow. Sometimes they shew particular faces 200 or 300 feet high.

By all accounts they are so full of rents and chasms as renders them exceedingly dangerous. The Ice too of which they are compos'd is less solid and transparent than common Ice, it being in reality frozen snow. The surface of the glaciers is not smooth, but rough and tuberculated, owing to its being frequently thaw'd and frozen again; and it is all of a blue colour. This I have seen in some of the mountains in the highlands of Scotland where the snow is perennial. But we know that frozen snow though not perennial assumes this blue colour and according to the length of time it continues frozen it becomes so much the deeper.

In consequence of this colour, we are told that when the Sun shines bright on a glacier, it exhibits a great variety of splendid prismatic colours, and when it thaws the melted snow water retains its blue colour, and all the river water in Switzerland which comes from the Alps in the summer season is of a sea green colour and retains this colour even at a very great distance from its source.

This colour is ascribed by Mr. Ray to nitrous particles existing in the snow water; this however is merely words.

In the Glaciers there are Avalanches. These are huge masses of frozen snow thrown from the mountains in the summer season, into the valley; these are occasioned chiefly by the heat of the earth itself melting and so undermining the snow at the bottom.[34]

By some, these immense quantities of snow in the glaciers are supposed to continue the same in bulk. The inhabitants of Switzerland however have an idea that they continue to increase gradually for 14 years, and during the ensuing 14 years they diminish in a corresponding way.

It does not appear however that they either increase or diminish with such regularity, but they certainly do not continue always the same, being altered by the particular seasons.

There are many causes which limit the increase of the glaciers, but chiefly the evaporation from their surface; the winds at such a height must be constant and violent, and we know that the evaporation from Ice and especially from snow is very considerable.

The next thing to be considered is the natural history of

ISLANDS

The Islands and Promontories over the whole globe extend in the direction of the neighbouring continent or in that of their own mountains; this obser-

[34] The effects of undermining and temperature in avalanche development are well described.

vation you will find confirmed in every part of the world. As in Britain, Madagascar and the West India Islands are all extended in the direction of their mountains. The great peninsulas of Caliphornia and Kamschatka, the three great promontories of Asia, Africa and America all extend in the direction of the mountains which pervade them, and that is generally always from South to North.

Nay even the smaller Islands always lie in the direction of the larger Islands to which they are adjacent as is the case with the small Islands about Britain, the Western Isles. There likewise and in every other place all contiguous Islands are similar in shape; thus where a cluster of Islands are grouped together there is one particular form to which they all approach. This is particularly remarkable in the Islands of the Kamschatka sea, and their form is indeed a very particular one. The largest of them which is Byrons Island is 150 Versts in length, the Russian Verst being equal to 3/4ths an English mile, but is in many places only from 3 to 6 in breadth and its greatest breadth only 23. There is no such disparity in the Islands of any other part of the world, and yet all these in the sea of Kamschatka are more or less similar to it.

In many Islands where there is a cape or promontory on the one side of the Island there is generally a bay in the other.

All distant Islands consist of high land, as the two distant Islands of Britain, the Pharo [Faroe] Islands and St. Kilda. Likewise St. Helena, the Island of Ascension, Juan Fernandez and Easter Island which are the most remote Islands in the globe have all high land.

Peninsulas that have a low and soft Isthmus may very easily become Islands, but we never hear of such actually taking place, on the contrary we have very legendary stories of remarkable disruptions in different countries; hence it was of old supposed that Sicily was torn from Italy, the Island of Cyprus from Syria, Asia from Europe at the Bosphorus, Europe from Africa at the Straits of Gibraltar, Britain from France at the British Channel; Cylon and Sumatra from the continent of India, and that Greenland at the one extremity and Terra del Fuego at the other had been divided from America. But why not America from Europe and Asia and indeed every one continent from another?[35]

These are all groundless and any suppositions, to which there is not the smallest foundation in history, tradition or nature, and yet some of them are credited and to this day affirmed by very considerable people. That the Thracian Bosphorus was once shut up, and that this was the cause of the Deucalion deluge was a very old opinion, and one supported by many of

[35] The concept of continental drift is old and still debated.

the Moderns especially by Tournefort,[36] and by the learned and accurate Pallas it is entertained to this day.

Of late also it has been and is now a very general opinion that all the lately discovered Islands in the South Sea have been originally connected with one another, and this is founded on a similarity of language prevailing among the inhabitants of these different Islands; but people seem here to forget that the Indians who have Canoes and that too of a very considerable size, may easily pass from one Island to another, and it is now well known that Canoes full of Indians of both sexes have frequently been discovered in the Pacific ocean, 40, 60 or even 100 leagues from any land. They have certainly made their way gradually from Asia, and this is confirmed by the similarity of their language with that of the Malais.

But likewise Islands from natural Causes become Peninsulas, and of this we have one remarkable and well authenticated instance in the Pharos of Alexandria, which was the light house of Egypt. We are well informed that at the time of the battle of Actium this light house was situated on an Island yet a considerable distance from the shore, but since that period it has been united to the Egyptian land by a tract, and this isthmus is formed evidently from the sediment of the Nile.

Another legendary story of this kind occurs in the writings of Plato. He had harboured some how or other that there once existed beyond the Hercules pillars, a very large island which he called Atlantis. To this story credit has likewise been given though it is unquestionably the very slightest of all ancient reports.

There is a set of Islands[37] formed in a very peculiar manner in the hotter climates and especially in the Pacific ocean called the

CORAL ISLANDS

The Coral shoots from the bottom from a very considerable depth and grows gradually upwards till it reaches the surface of the sea. It there gathers the exuviae of the ocean, and the seeds of plants. It spreads gradually all around, acquires soil, and in this way there can be a regular progression traced between a single plant of Coral and an extensive Island filled with plants and Animals. In some of the Islands of the Pacific ocean there is high

[36] Joseph Pitton de Tournefort (1656–1708), professor of botany at the Jardin du Roi.

[37] This short discussion on coral islands is worth special attention. In a few words he describes vividly and accurately the essential features in the growth of a coral island. He takes the reported observations of voyagers and reconstructs the proper sequence of events in what apparently is the first statement of a geologist on the growth of reefs.

land of primitive rock, but then to the shores of this high land the coral has attached itself, and by spreading all around formed an extensive plain which the sailors call a Coral reef, and in consequence of this a plain soil is formed. Hence in most of these Islands as in Otaheite [Tahiti] there is a considerable plain between the mountainous part and the shore, which is evidently formed in this way.

It is remarkable however that notwithstanding the many accounts we have received of these Islands from the different voyagers, yet we do not find the particular species of Coral mentioned by any of them; for though many corals possess this property of spreading, yet there are undoubtedly some one or more particular species which are best fitted for this.

We come now to consider in a particular manner the

STRATA

of the Earth,[38] and here as in the other parts of Geology, I would not wish to be thought to deliver any thing like Theory, but merely a natural history of the earth.

The strata, veins and fissures over the whole globe are every where so extremely similar that they evidently must have been formed by one general and uniform Cause, not by causes partial and irregular as earthquakes and volcanoes.

The Strata of the globe bear all the appearance of a subsidence of a fluid; from a fluid we know that solid bodies naturally subside, stratum super-stratum; thus we may see every day in the bed of the ocean and rivers, strata of sand, clay, gravel etc. disposed exactly in the same manner and on the same principles with the Strata over the whole globe.[39]

The most ancient Strata of the earth are either those at a very great depth or on the summit of the highest mountains.

As for the strata of earthy matter it is mostly confined to the surface, and is spread over the primitive strata, often over the secondary, but most frequently among the Accidental strata.

The Strata of plains are more [horizontal?] and more regularly disposed than those of Mountainous countries, and they are likewise generally more extensive.

---

[38] As a whole this is a splendid section and deserves the thoughtful attention of the reader. Walker has assembled in these pages almost everything known at that time about rock strata and makes many interesting deductions.

[39] Every geologist will recognize this as a remarkably clear statement on the origin of sediments, their transportation, and deposition as "stratum superstratum."

It is pretty clear that the Strata of the earth are not disposed always according to their specific gravity, and yet the whole of Dr. Woodwards theory of the Earth hangs on this affirmation. (He supposes all the strata were dissolved or diffused in a fluid, and that they subsided according to their specific gravity.) For at the depth of 100 fathoms we frequently find fossils much lighter than those near the surface. In general however the deepest strata of the same materials are the most solid and compact, but this appears to be a necessary and natural consequence of the compression to which they are subjected.[40]

The inclination of the earth's surface every where depends greatly on the inclination of the strata below; where the strata are horizontal or but little elevated there the country is flat; on the other hand where the strata are vertical, or highly elevated, there of consequence the country is steep.[41]

The Streik is a longitudinal direction of the Strata, and the Dip is the inclination of the stratum to the horizon.[42]

I may here still further consider the distinctive characters of the primary, secondary, and tertiary or Accidental strata.

I have already observed that the primitive strata are always vertical, that is disposed at an angle with the horizon of from generally 60° to near 90°. They are never found horizontal excepting in one case, that is when the horizontal alternate with the vertical strata; for example all the Basaltes are vertical, but in the case of the greatest collection of Basaltes we are acquainted with, the Giants Causeway in Ireland there [are] immense quantities of these vertical columns all grounded on a horizontal base, or placed upon a different rock. This you may observe likewise in some degree in Salsbury Hill in this neighbourhood; there the whin rock is vertical, but in several places you will find them placed on strata not indeed horizontal but much inclined to the horizon.[43]

This is an appearance but little known, and not taken notice of by any of the Theorists of the earth, for it is generally assumed by them that the Primitive strata are never horizontal, but they have only seen them in their vertical position; if the base of the Salsbury rock was shut up and hid from the sight every body would imagine it was uniformly a vertical stratum.

[40] This is one of the earliest statements on the increase of compactness of sediments under compression of the overlying load.

[41] The relationship between topography and structure is recognized here.

[42] This sentence with its excellent definitions of strike and dip is from D.C. 2-18.

[43] An angular unconformity is described. Walker has not interpreted the meaning of the unconformity, but he has clearly observed the relationship of tilted strata underlying a horizontal bed.

But the whole body of these rocks may be considered as one stratum, divided into horizontal and vertical masses.

With respect to the Secondary Strata, they are all horizontal, or inclined at the smaller angles of 20°, 30°, or 40°.

The Secondary strata generally occupy the sides of Mountains. They extend themselves about the middle and towards the foot of the Mountain; but they chiefly occupy valleys and plains and accordingly in general, in Scotland all the highlands and mountains and all the high land are composed of primitive strata. On the other hand what little level country we have and the valleys are all composed of Secondary strata.

Of all the Secondary strata in Britain that of Chalk seems to be the most extensive and the deepest. I remember once to have gone into a great Chalk quarry in Berkshire where one had opened to the eye 90 feet deep of chalk. I asked the quarrier if ever they had reached the bottom; he replied very gravely it had no bottom, but that the earth consisted of Chalk. I made no answer, but I thought his Theory was just as good as that of many others.

In one place I found a pit well dug in the Chalk down to the depth of 230 feet, and indeed in the Midland counties of England where chalk abounds they no where can find bottom.

The Chalk in Britain and in France is every where a superficial stratum, it being never placed below Lapideous strata,[44] the same is the case with gypsum, the substance of which plaster of Paris is made; the gypseous strata are all superficial. If they happen to have any other above them it is merely earthy.

There is one stratum of the Globe whose origin is a very difficult problem in mineralogy, that is the strata of fossil salt. There are immense masses of these in Poland, Germany, Spain and even in some parts of England as in Cheshire. There is one appearance however which shews them to be all secondary strata. The late Dr. M—— who made many careful inquiries on the subject when he visited the great salt mines at —— in Poland, found them digging out a ships keel of a very considerable size from the depth of about 50 fathoms, and this plainly shews that they are to be considered as secondary strata.

As the petrified plants and Animals which are found do all subsist in the secondary strata, it has therefore been concluded that the Primitive strata were in point of existence previous to all organiz'd bodies. But that on the contrary the earth had been inhabited both by plants and Animals before

---

[44] This comment shows that Walker failed to recognize that chalk deposits were interstratified with other sediments.

the formation of the Secondary Strata is evident, as in these strata we find their remains.[45]

The Accidental Strata occupy generally the deep valleys the extensive plains and the sea shores; they are very copious likewise on the banks and also at the mouths of rivers, especially at those of great rivers. There they are found in great quantity and to considerable depth, being the sediment of the rivers.

The famous geographer ——— was at the pains to examine the different strata in Holland. On digging a well 231 feet deep, they dug to this depth into the tertiary strata which consisted of strata of sea clay, gravel, sea sleech etc., and it is highly probable that from the same cause . . . the depth of the Accidental strata must be much greater in the Lower Egypt.

There is one thing pretty certain, that all rocks containing the remains of organiz'd bodies of plants or Animals certainly cannot be considered the produce of Volcanoes.

The Secondary and Accidental strata, especially the latter, form the countries most remarkable for their fertility, as the lower Egypt, Holland, likewise some of the low and fenny parts of England. And the cars[46] countries as we call them, in our own country all consist of Accidental strata, and always offer the most perfect soil.

At Boroughbridge a whitish limestone becomes the general stratum under the staple, which is usually from one to three or four feet deep. The upper part of stratum is a soft cliffery rock, with which the road is made. It breaks down and cakes exactly like soft whinstone. This limestone continues to Doncaster. Between Doncaster and Foston in Lincolnshire, this limestone disappears; at least it lies not anywhere so near the surface as to be worked. The soil and the road are sandy, and deep strata of sand are to be observed in some places, mixed with gravel.[47]

[45] Walker says little about petrified fossil plants and animals, but in this paragraph he is almost prophetic when he states that life must have existed on earth when the primitive strata were decomposed, simply because we have the record of them in the overlying secondary strata. This argument has been used continuously until recent times and the finding of actual fossils in presecondary strata.

[46] Carrs or kerss—low portions of valleys, including floodplain areas and terraces.

[47] This paragraph and the six corollaries are introduced from Walker's "Mineralogical Journal from Edinburgh to London" (*Essays*, pp. 399–400). It was probably written in 1769 and certainly not later than 1770. The section is included to show something of Walker's field observations and his practice of correlating rock units from place to place.

A line from Boroughbridge to Doncaster parallels the strike of the Permo-Triassic. Walker was probably following one of the Permian limestones along this strike. Boroughbridge lies approximately 40 miles north of Doncaster and so Walker had followed a single unit for a considerable distance, correlating it from outcrop to outcrop. He says that it disappears between

At Foston, again, the same general stratum of whitish limestone appears, and continues in the same manner as before till within five miles of Stilton.

Corollaries:

1. All the strata in a flat country are horizontal or nearly so.
2. The strata are more extensive, uniform and unbroken, in a flat, than in a hilly country.
3. The staple is always deepest in a flat country, where the strata are horizontal and extensive; except where chalk and flint gravel prevail.
4. There are no limits between chalk and limestone. They are only gradations of the same stone.
5. Chalk often contains a horizontal stratum of flint in loose masses, from one to fifty pounds weight as in the Irish limestone. Each stratum has only the thickness of one of these masses.
6. Sand, clay, gravel, sandstone, limestone, coal, chalk and flint, are the proper fossils of a flat extended country, and the only ones to be observed on the high road between Edinburgh and London.

Observations:[48] Marle is found only in mosses and boggy places where formerly there may have been a stagnation of water, the natural habitation of these testaceous animals. In all mosses, the stratum of marle is deepest in the middle, and grows gradually thinner towards the sides.

This observation tends to prove marle to have originated from the shells of snails, which in a pool of water, would be spread thickest in the middle, where the water is deepest.

There is sometimes a stratum of moss, included in a stratum of marle. Particularly, on one place, where the marle reached the surface, it was a foot and a half deep, beneath which there were fourteen inches of pure moss, and under that another stratum of marle.

Of this appearance we cannot as yet presume to give a satisfactory account.

Doncaster and Foston: quite true, because it dips below the surface. He says it reappears at Foston. Unfortunately, at Foston he is higher in the section and is now dealing with a Jurassic limestone, and so his last statement is false.

In his description of limestone he refers to the Sutton limestone. Sutton lies about five miles north of Doncaster and, according to the geological map of England, rests on or near the Upper Permian marl or the Upper Permian magnesium limestone. This is certainly correlation by physical means and represents also the naming of a rock unit by use of a place name.

[48] The remainder of this section is introduced from Walker's *Essays,* "History of Shell Marle" (Edinburgh, 1808), pp. 313–22. It is included to show Walker's approach to the problem of finding the origin of a formation. Here moss is synonymous with peat.

As it is here delivered we can aver it be a certain fact; and on this subject it seems to be the *Observatio Crucis.*

It may lead us to think that the stratum of marle could be formed above the moss in no other way than by means of shells.

On this supposition, we might expect to reap a growth of marle from a plantation of these animals.

Experiment: When marle is long calcined, it grows black, but the shells that are in it, though likewise calcined, continue white.

Corollary: This shows, that the marle, and the substance of these shells, are of a very different nature.

Observations: Limestone, marble, chalk, and some earth minerals, are all alkaline, and frequently filled with shells; yet none ever imagined that they had any such origin as that above ascribed to marle.

Beneath four feet of moss, we knew once a bed of grey marle, three feet thick, and full of shells. Under this, there was a stratum of clay and stones, a foot thick, and beneath that, two feet more of pure white marle, which had no shells.

It is a frequent case to fall in with a band of clay, in the midst of a bed of marle. The marle above the clay is full of shells, but none are to be found in the clay, or in the marle under it, so far as I have observed.

This is a presumption, that the marle under the clay, has another origin than shells, and is a native *sui generis fossile.*

Some think, that where these shells abound in stagnating waters, they are a sure sign of marle; but this is quite a mistake. These shells are to be found in abundance in the stagnating waters of almost every part of Scotland, and in many places where, upon repeated trials, no marle could be found.

We next go on to consider the

VEINS *and* FISSURES

found in these Strata. The form of Veins in these rocks or strata seems very analogous to the form of fissures taking place in any solid body in an indurated state, that they are similar to the fractures in solid bodies of great bulk produced by partial subsidence.

It would appear however that these veins in strata, have taken place since their induration as it is highly probable that all the rocks at a former period were in a fluid or very soft state.

Veins are strictly speaking those fractures which run thro' and join a number of contiguous strata. On the other hand Fissures are those fractures which are confined to one particular stratum.[49] A third kind of fracture is that commonly called by Miners an Intersector; this is a transverse separation extending from one vein to another.

All the fractures occurring in strata may be referred to one or other of these three kinds, Veins, Fissures and Intersectors.

In consequence of this all the earthy strata are totally devoid of veins or fissures, and yet it appears very probable that if these were sufficiently indurated, then fractures would be formed.

Those strata that are horizontal, never have in them what miners call Veins, but only either perpendicular or horizontal fissures.

Veins are formed only in strata of a vertical or considerably elevated position. In consequence of this they are always more frequent in primitive than in secondary strata. We have some particular species of rock entirely destitute of veins or fissures, as the coarse sand stone, called in this country whin stone.[50] Such also is the coarse rock called the coarse pudding stone, and such is the rock or stone called by naturalists the *Sara coacervata* which are found heaped together in one mass and without any veins or fissures.

It appears in general that the veins, fissures and Intersectors are more numerous near the surface than at a greater depth; so far at least as man has penetrated into the earth, this seems to be the case, or is at least found to be so to the depth of 140 or 150 fathoms.

The fractures of the Globe are the canals by which the subterranean waters are convey'd; these fractures are generally found filled with earthy or Lapideous matter, very frequently with metals and with a great variety of ores. But where fossil matter is thus found we may be allowed to form the conclusion, that it has either been dissolved or diffused in water, that it must have been carried from the adjacent walls of the fracture and lodged in its cavity and hence the spars or ores found in these fractures must have been dissolved or diffused in water; and hence all such are called of a venigenous nature or venigenous fossils. Some of these found in veins, have evidently been derived

[49] The idea that fracturing was caused by partial subsidence and that sediments were originally deposited in a fluid or were in a soft state before induration are indeed "modern." We have only refined and built upon the concepts. Intersectors refer to cross veins connecting two or more other veins.

[50] Walker knew that sandstone was not whinstone. He describes both on many occasions without confusion. I believe that Walker either originally wrote or meant to write, "as the coarse sand stone, and what is called in this country whin stone. . . ."
In another manuscript version (D.C. 2-18) Walker adds at this point: "Some philosophers have given very ingenious suppositions and different accounts of Veins and Fissures and concerning this formation; but I believe the simplest of these is the best. It is supposed that all strata appear to have been at one period or other of some soft consistence. Veins and Fissures therefore are the cracks and crevices, the natural consequence of induration taking place."

from the walls of the vein itself and are of the same substance, yet they are generally harder, more compact, and of a finer grain than the strata themselves, having been either dissolved or most minutely diffused in water.[51]

The veins in mountains generally run thro' the mountainous ridge obliquely. There is no vein yet discovered exactly placed at right angles to the horizon. Some however approach very near to it and they are sometimes found at an angle of 80° and sometimes even at one of 85°, but still no body I believe has even seen one placed at right angles to the horizon.[52]

It is observed by miners that almost all veins as they go deeper gradually approach to a greater angle than they have near the surface, and it is likewise observed by them that the more veins approach to a perpendicular they are always the wider, and if they be metallic they are consequently always richer.

The veins in different parts of their course are very variously contracted and dilated. A vein in one place comes gradually to contract, and again to widen. I have seen a vein of Lead in one place, bellying out to six feet in breadth, and within the breadth of this room becomes so contracted that they could not insert the blade of a knife betwixt its walls. But the miners are by no means discouraged at this contraction if it has only the two walls distinct, they go on and in a little space come again to perhaps six feet of Lead. When a vein contracts the miners say it c———, and when it widens they say she bellies.

When the walls of a vein are composed of hard materials, it is always narrow; but if soft it is always widest.

I have only one thing more to observe on this subject and that is: Veins are sometimes interrupted in their course; they are sometimes entirely cut off by it; at others there comes a body of a different fossil across the vein placed at right angles to it, but on the other side the vein continues as before, as happens in the mines of the Isla. There the metallic veins are embody'd in Limestone. But there in many places what the miners call whin dykes, or masses of whin stone, from 1 to 9 feet in thickness . . . come across the vein. But on the other side of these the veins observe the same direction and the same declination as before.[53]

The veins in the earth too are very cavernous, containing considerable vacuities, these are called by the Miners [vug] holes and are generally filled

[51] In this paragraph the movement of subsurface water is used to explain the solution of rocks, the transportation of the dissolved or diffused materials, their deposition in veins and cavities, and the lateral secretion of vein material.

[52] A few vertical veins do exist. In speaking of dip angles, Walker overlooks the law of probability.

[53] He is describing a vein cut by a dike.

in metallic veins with the most beautiful Chrystalization, and it is noticed by miners that such veins as have [vug] holes are always the richest in metals.

I come now to take notice of the

Caves, Grottos *and* Caverns

occurring in the strata of the earth.

What we call Caves are excavations in the rocky strata, most frequently formed on the sea shore or banks of rivers.

Some particular species of rocks are better adapted for the formation of these than others; those especially which are disposed in thin strata, which are easily disjointed by the waves. Such likewise is the rocky strata in the Islands of Isla and Jura called the ——— Jurae. It is extremely fissile,[54] most parts of both Islands are composed of it, and the force of the waves demolishes it to such a degree as to form very capacious caves.

The common schistus or Slate is likewise exceedingly well adapted for the formation of caves, and accordingly they abound in those of the Western Isles which are composed in great part of Slate.

There are however caves formed in other rocks, especially from those of a soft mouldering sand stone. It is in a stone of this kind of which the magnificent caves of Forfarshire, Kincardineshire, and in the Isle of Bute are formed.

What are called Grottos are those which ———. These most frequently occur in mountainous countries. There is one described by Scheuchzer near the summit of Mount ——— of an unknown depth. They most frequently occur in Calcareous and Gypseous strata, because Limestone of all others unless Gypsum is the most easily abraded[55] by water.

Dr. Pallas describes caverns of prodigious depth in the Uralian mountains, but then in that part of the world the caverns are all formed of Limestone.

On the banks of the river Silva in the Russian Empire, he describes some immense Grottos and subterranean passages extending for about five English miles in length and 2 in breadth, and the whole excavated from a mountain of [limestone?].

We have in Britain two very remarkable subterraneous caverns, one of them in Glocestershire which is called by the inhabitants Penpark hole.[56] The

[54] Capable of being divided or split.

[55] It is not clear whether he includes solution in this word.

[56] Pen Park Hole is near Bristol, about 50 yards north of the junction of Pen Park Road and Lanercost Road. The cave is developed in Carboniferous limestone near a north-south fault. The top of the main chamber was opened about 1775. It is 350 feet across and 200 feet deep. A pamphlet (*ca.* 1775) on Pen Park Hole is in the Bristol Reference Library. (Personal communication from Derek Ager, Department of Geology, Imperial College, London.)

passage down to it is 177 feet deep. It is oblique and very steep, from six to 12 feet in breadth, but in such a situation that people with some difficulty can scramble up and down. At this great [depth?] we come into a subterraneous cavern 200–300 feet long, 100 broad and 50 high. There is in the middle of it a pool of water 80 feet broad and at sometimes 16 feet deep. But the Gentleman who describes it went there in the month of September when it was very dry, but at other seasons he says it rises above 18 feet additional to what he saw it, which plainly shews it to be filled entirely with the surface water.

The other remarkable cavern in England is in Derbyshire and called Elden hole. It is a perpendicular pit the depth of which has never yet been discovered, for altho sounded by a line of 884 yards in length, the plummet still drew.

Mr. Maupertuis long ago was fond of the enterprise of a philosophical pit being dug for philosophical observations. This of Elden hole is indeed made by nature, and is such a one as we could not expect from art because notwithstanding its depth it is perfectly dry. And indeed that its depth has not been explored is no encomium on the present age of enquiry.

We come next to consider some of the factitious strata of the globe, and first the common

STAPPLE

Which is the Mould or vegetable soil;[57] what Pliny very properly calls the *cutis terrae* and which by some other authors is called the *corium telluris*.

This is that stratum which every where covers the earth and which serves every where as the Matrix of vegetables. This evidently is a factitious stratum, and the produce of time. You see it every where very distinct from the subjacent stratum.[58] It is sometimes only a few inches, sometimes a few feet, and in some cases some yards in depth.

As it is every where the general reservoir of every kind of matter it must necessarily vary very considerably as to its contents; but the true Stapple strictly speaking consists only of three essential ingredients, and these are clay, sand, and vegetable and Animal matters. No other articles are necessary for the formation of a perfect soil. You know likewise it varies exceedingly in different parts in respect of Colour, being sometimes brown, some-

[57] We spell it staple—a deposit of decomposed vegetable soil overlying bedrock.

[58] Walker realizes that soils are a product of time, that their black color is derived from organic substances, and that they are a product, in part, of the underlying substratum as developed through the subsoil. These concepts are more fully brought out in the essays on peat and on shell marl.

times grey, red, yellowish but in general it approaches to blackness, and the blacker it is usually always the richer. It derives its blackness from Animal and Vegetable matter, for all Animal and Vegetable substances when turning into putrefaction assume a black colour, and accordingly those lands which have been immemorially dunged, as those in the neighbourhood of great Cities, are always blacker than those which have been less manured, and even upon a single farm we may see the effect of manure. In every farm there is a part called the infield which is generally dunged every year. The soil of it is always very black, whereas that of the rest of the farm is perhaps brown or reddish.

The Stapple is either thin or thick, very much according to the subsoil or stratum immediately below. It is generally thin when it lies on sand, gravel, chalk, or rock; on the other hand it is generally deep when it lies on clay. The reason is plainly this, because the sand, gravel, chalk or rock, are exceedingly pervious, and hence the rain or flood water immediately sucks in them to a great depth, but the clay resists the [flow?] of water, and therefore the water communicates an additional quantity of Sediment to the soil. Hence the soil is always deeper in plains than in Mountains, and is generally thickest in deep valleys where there is a great deal of water collected, which stagnates and leaves a copious sediment.

The Stapple likewise always partakes considerably of the subsoil, and hence we have either a sandy, a chalky or gravelly soil.

There is another appearance here to be taken notice of, those large beds and banks of Gravel consisting of small rounded stones, and likewise those large round masses found in almost every country over the whole globe, [lying] upon or near the surface called with us

Bowlder Stones

and by Naturalists *Noduli Vagi.*[59]

These are of very different quantities, and found in a dispersed state unmix'd with any fix'd rock. Sometimes they are found of a prodigious size but always somewhat of a convex or rounded figure.

We know that when fossils are round, they consist of concentric layers but this is not the case with these Bowlder stones, which are always perfectly solid.

[59] I believe this is the finest description of glacial drift up to Walker's day. He does everything but identify it as glacial in origin. According to his own statements, he searched for many years for the sources of some of the pebbles making up the drift and was rewarded only in discovering the origin of some flesh-colored jasper. He makes many accurate observations on drift.

They are to be found in size from that of a pin head to enormous masses, the moor stone that covers all counties is of this kind. We have in Scotland abundance of these, some of them consisting of Whin rock, and some of granite. That immense mass of granite brought by the Empress of Russia to Petersburgh for a pedestal to the statue of Peter the Great was a mere rounded nodule.

These *noduli vagi* are very various in their composition, but consist always of primitive rock. They are found chiefly on the surface of the soil and never at any great depth, they are never found in the primitive rocks, but are frequently found in a sporadic state lodged in the secondary strata, as in Limestone, freestone, millstone, etc.

They are never found at any great height on the mountains. There are often indeed found at a great height on mountains loose rocky masses but these are always angular and never rounded and they are evidently of a very different and posterior origin to the Bowlder stones. On most of the mountains of Scotland you will find loose masses of a rocky nature as on Salisbury Craggs but all those are angulated.

It is remarkable that these vague nodules, when of a great size too, are found where no fix'd rock of that kind is known to subsist; they are likewise generally composed of those rocks which break not into round but angular fragments.

One thing is clear that they never have been formed where they are now found, but elsewhere and at a very great distance.

They all of them too belong to fix'd rocks existing in some part of the country in which they are found. Thus there is a flesh colour'd jasper found in nodules in Tweeddale, Nithsdale, and Anandale, and yet it was long before I could find any fix'd rock of that kind. I found it last however in great abundance on the Tweed.

It is evident that these vague nodules are water worn or have receiv'd their rounded figure by attrition in water, like those which are found by the sides of rivers and on the sea shore which are rounded exactly in the same way with the Bowlder stones.

But it is clear that the rounding of such nodules could never be affected by any temporary current of water however violent, but they must have acquired it in a great length of time.

At what Æra or by what agent these vague nodules have been torn and transported from their original fix'd state to their present situation is as yet an insoluble problem.

They have not even been much noticed by Naturalists till very lately, but of late years they have been mentioned by several writers, and each of these has formed an opinion concerning their origin. Thus the Abbè Fortisse[60] thinks that all of them have been formed and deposited by Ancient rivers which now no longer exist, but this instead of explaining their origin is a mere [conjecture?].

By Mr. Ferber they are imagined to have been deposited by the present rivers which he supposes to have run at one period[61] 1000 feet or upwards above their present level; but this is begging what we by no means can grant, and which even if we did seems insufficient to the effect.

Mr. Whitehurst[62] likewise describes them, and thinks that all of them have been thrown out by subterraneous explosions. Mr. Kirwan thinks they have been all projected by Volcanoes.

But these different opinions are certainly quite incompatible with the nature and properties of these nodules as they have been now described.

They certainly constitute a Phenomenon, one of the most curious and to me at least one of the most difficult to be accounted for, in the natural history of the globe. When you find a rounded mass of granite or whin stone, 10 or 12 tons or upwards in weight thus rounded in water ... if I could be certainly informed when or by what agent such nodules have been placed where they are now found, I think I could then and not till then give some light into the theory of this earth.

We come now to notice another factitious stratum

SEA SAND

Linnæus had a very peculiar opinion concerning its origin. He thinks that all of it is formed in the ocean by a combination of the earth of Rain water, with the *æthereum nitrosum* as he calls it, but this is a mere visionary opinion and this *æthereum nitrosum* spoken of by him and other naturalists is scarce an entity.

It is commonly supposed that a great part of our rocks is composed of Sea sand gradually indurated in the course of time. But it is rather more probable that the sea sand is derived from the rocks than rocks from the sand, and accordingly the sea sand all the world over always partakes of the nature of the adjacent rocks. Thus in the Islands of Isla and Jura there is that remarkable stone I mentioned before, the greater part of which consists of a whitish

---

[60] See note 21.

[61] The word is used here in a temporal sense.

[62] John Whitehurst, *An Inquiry into the Original State and Formation of the Earth* (London, 1778).

Quartz, and accordingly all the sea sand on the shores of these Islands is pure white in so much that it has been used for the manufactory of Chrystal.

In like manner on the Island of Lismore and in the Long Isle, the shores are every way covered with black sand, the rocks on these Islands consisting of black and blue slate.[63]

The Sea Sand in several parts of the world is a great nuisance to the inhabitants by means of its formidable invasions on the adjacent country. This is well known in Holland, likewise in several parts of England, and with us too especially in Aberdeenshire. In one place on the Suffolk coast there was a remarkable devastation took place some time ago and the sand was carried to a great distance. The sand hills only occupied about 8 acres but the sand rendered 1000 acres of excellent soil entirely useless.

The best remedy for this destructive property of Sand is immediately to cover the surface with such plants as naturally grow in loose sand; there is a plant grows in that situation in the Western Isles, the Yellow Ladies Bed Straw (*Galium verum* Lin.) and accordingly it is so well fitted for fixing the sand that in the Hebrides a severe penalty is annex'd to the pulling it up; as is likewise the case in Holland to the pulling a root of the *Arundo Arenaria.* ..

But the best plant for securing the sand is the Spurry (*Spergula arvensis* Lin.), which is fitted for growing in blowing sand. It falls down in winter into a sort of slime and in a few years form a kind of soil.

We shall now take notice of

QUICK SANDS

These so far as I have noticed, are of two kinds.[64]

One on the sandy shores where there are springs which rise within flood mark. They do not rise to a head and run in a continued stream, but spread themselves under the sand, incorporate themselves with it and thus become very dangerous.

The other kind of Quick sands happens where there are springs in a clay or sleechy soil within sea mark. These make the clay very thin and become exceedingly dangerous both for man and beast; the quick sands of the river Isk on the border between England and Scotland are of this kind. But they

[63] In another manuscript version (D.C. 2-18) Walker states, "I am of the opinion, that all sand is afforded by the wearing and attrition of rocks." Such thoughts, when linked with his knowledge of the origin of deltas, the world distribution of strata, the origin of chalk, and other related ideas, show that Walker had a clear conception of the processes of erosion, transportation, and deposition.

[64] A search has failed to uncover any older explanation of the origin of quicksand. Walker's deductions were made after field observations.

are exceedingly changeable being sometimes quite solid. This is entirely owing to the banks of Clay being laid as it were above the level of the spring, but in a few tides after being again levelled it becomes a quick sand.

The next artificial stratum to consider is

Peat *or* Turf

Concerning which, and the origin there have been a great variety of opinions. —

Dr. ——— a Dutch writer wrote a treatise expressly on the subject, but his observations seem to have been entirely confined to that sort of Peat found in Holland. Accordingly he describes it as being composed of putrid animal matter, and that for a very bad reason, for he says that putrefied wood loses its inflammability. But we know this is not the case, and the Turf of Holland is very different from that of other countries. It is lighter, and is composed of putrefied animal matters and aquatic plants.

Scheuchzer was of opinion that Peat was a bituminous earth mix'd along with vegetable substances. There is not however any peculiar bitumen found in the Peat of this country. But in some parts of England we know this is the case.

In several parts of the world, among the vulgar, Turf has been considered as a submarine stratum; this is an opinion prevalent among the vulgar in Holland because on that coast great masses of Peat are frequently thrown out of the Sea, and in the harbours it is frequently dragg'd up by the Anchors. But this does not prove that Peat has been formed at the bottom of the sea, it only serves to shew that the sea has made great encroachments on the coast of Holland. The existence of the Roman fort of ——— is an evidence of this, which is now 2 or 3 miles [offshore] under water but which was formerly as far from the shore.

But that opinion has prevailed even with the vulgar of this country. There is an extensive Moss on the borders of Perthshire and Stirlingshire called Moss Flanders and it is the opinion of the inhabitants that this Moss has been wafted by the sea to the place where it now exists.

But in order to acquire a proper knowledge of its origin, let us endeavour to enumerate its nature and properties; for this purpose its different kinds are first to be noticed. These are as follows.

One kind of Peat or Turf evidently consists entirely of putrid wood or arboreous plants.

The second species is composed entirely of aquatic substances, half putrid and decay'd aquatic and Ulvaceous plants. We have some of this kind in Scotland and it is the most common in Holland.

The third kind is called Mountain peat. It is found only at a considerable height on Mountains; it never forms a thick Stratum, but sufficiently so for the purposes of fuel. It is evidently formed entirely by the decay of one plant, the common heather, the *Erica vulgaris;* and accordingly you find the roots and stumps of heather in this kind but never the remains of any arboreous plant.

Another kind is that called the flow moss. This is the lightest of all and is formed entirely of the Musci and particularly of that kind called *Sphagnum palustre.* There is one plant in this country which forms it in great abundance the *Hypnum f[luitans]* of Linnæus. It grows in water, and its growth is so quick that in the course of one year it will form from half an inch to a whole inch of Moss.

All these four kinds of Turf are frequently to be found mix'd with one another, but in all of them their composition and structure are always observable to the naked eye but then indeed they vary exceedingly. Some too are so thoroughly consumed that their regular structure disappears, as is the case with many peat mosses in the lower and warmer parts of Scotland, where from the greater warmth the putrefaction has been carried on farther than in the Mountainous parts of the Country. In short it is sometimes reduced to a mere earthy matter and in this situation the inhabitants cannot cut it into oblong squares for the purposes of fuel, but are forced to knead it with water and then dry it.

Peat is very liable to be carry'd off by water from high to low situations, and it then in the course of time forms a very deep stratum, called with us Water Slain Moss and is never so good fuel as the other kinds.

It would appear that all peat mosses have been originally formed in those places where we now find them.

There are many reasons occur in this country in support of this opinion. For example in the low countries the forest trees are always higher than in the mountainous parts. Now in all the low parts of the country where peat mosses exist the Trees found in them are very large, whereas those found in mosses in the mountainous parts are comparatively small. It is likewise to be noticed in Scotland that in the lower parts where the oak naturally grows, there the prevailing wood in mosses is the oak. On the other hand in the mountainous parts the tree most commonly found in the peat moss is the birch which naturally grows in such situations.

The Scotch fir or pine is found in great abundance in many of the peat mosses in the highlands, but no where in those of the South of Scotland. I have never heard of its having been found South of Tay.

There have been at different times found in the peat mosses of Scotland various remains of antiquity lodged at a great depth and these have been thought to furnish a presumption that the whole depth of moss found above them has been deposited since they came there. For example a Roman Camp Kettle was found some time ago lodged 6 feet deep in Flanders moss. It was very large. The Romans were unacquainted with cast iron and they used brass extremely thin. This kettle was a fine piece of workmanship and I suppose it would hold 10 English gallons or upwards.

In the deep Lochart Moss near Dumfries was found 6 feet deep a leather bag filled with Saxon coins and a silver crucifix, and it is well known that the Saxon Heptarchy reached to that place.

A few years ago a beautiful piece of Roman antiquity was found in Anan moss in Anandale only three feet deep; it was a ———— of pure gold designed probably for the head of a Standard to some principal officer and there was an inscription on it *Jovi Augusto*.

These and similar pieces of antiquity are frequently found in peat mosses and the usual conclusion drawn from them is that the superincumbent moss in the course of years and ages has been formed above them; but then they have been probably hid in the Mosses.

But there are other remains of antiquity to which this objection does not lie, as the horns of deer which are frequently found at the very bottom of peat mosses in Scotland. The horns of the Elk have been found in such situations in this country, an animal which was certainly indigenous in Scotland but of which we have no account either in history or tradition unless what these horns afford. It is probable that when the country was full of Red deer and the other larger species, which is now lost or unknown, their horns behoved to drop annually, and it is certain that the moss superincumbent to them has grown after their deposition.

The deepest mosses are in the deepest low situations owing to the more luxuriant vegetation.

It appears in general that most of the peat mosses in Scotland have been formed where they are now found, by the decay of Ancient woods, and hence the moss timber on those trees found in mosses are generally divested of their bark, the trees having stood till they have lost their bark, and they generally lie in one direction with their heads to the North East which shews that they have been blown down by our prevailing winds from the West and South West.

On the other hand in Holland the Moss trees are commonly found with their heads directed to the south east, because in that country the strongest winds blow from the North West.

There was long ago an old Gentleman in this country George Earl of Cromarty who lived to the age of 90 or upwards. He gives a very curious account of the rise and progress of peat moss in the north of Scotland, in a letter to Sir Hans Sloane which was published in the Philosophical Transactions for the year 1711. When he was 19 years old he was with his Father in West Ross on the estate of Cromarty in the year 1651. There he saw an extensive wood of Scotch fir, very old, and standing without its bark. Fifteen years afterwards being in the same place he found this wood all blown down and both the trees and all the surface of the ground covered with green moss. But in the year 1699 happening again to visit that part of the country he found his tenants digging their peats in that very spot where the trees had stood, and the substance of these trees formed the principal part of the moss.

Dr. Morton[65] informs us that there have been fir trees found in the deep bogs of Northamptonshire, and it has been concluded that these trees have been brought from a distance and deposited there, as the Scotch fir is not now a native of England. But it is by no means improbable that the Scotch pine may formerly have been a native of England. The authority of Caesar is urged against this. This great man made many excellent observations wherever he went, and even in the short stay he made in the south of Britain. He says that there is in England every tree that grows in Gaul *"præter Fagum et Abietem."*

The Abies of Caesar has been thought the Scotch pine, but it is more probable that Caesar was better acquainted with the difference between the Abies or spruce fir and the Scotch pine, and he certainly knew them better than many of the Moderns seem to have done. The Abies indeed never grew naturally in England but it is very likely the Scotch pine had, and it is found in several of the bogs in England.

There is likewise a surprizing similarity in the depth and extent of peat mosses in different parts of the world where they exist, which shews that they have been all nearly coæval. In general peat mosses are found from 5 up to 20 feet in depth all the world over. In Ireland within this few years peat has been dug 20 feet deep. In Scotland our deepest mosses are generally about 12 feet. I have indeed seen them 18 feet deep but this more rarely happens.

[65] John Morton, *The Natural History of Northamptonshire.*

In the Falkland Islands it is likewise generally about 12 feet. In Holland it is frequently of that depth, but often much deeper but then it has probably been carried and deposited there by the sea.

Peat is the peculiar product of the temperate and cold climates. It cannot possibly subsist in a warm climate. All peat is formed of half putrid Vegetables, but putrefaction is well known to be a species of fermentation, to which Animal and Vegetable substances are subjected and it is slow or rapid according to the degree of heat; with a considerable degree of heat it proceeds rapidly and brings the Animal or Vegetable substance to the state of a perfectly inert earth in a short time; but with only a small degree of heat it proceeds more slowly, and this is the case with the Peat mosses.

I have never been able [to] fix the limits of peat moss over the globe; but I am inclined to think that it does not exist within 35° of the line, and certainly not within the Tropics. Accordingly all the peat in the south of England is much more putrid or farther advanced in the putrefactive process than that in the North of Scotland; and the peat mosses in the low country of Scotland are even more putrid than those of the high and elevated parts of the country or on a mountain owing to the greater degree of warmth. There is another remarkable instance of this too in the Isle of Lismore one of the Hebrides. I never in my life saw the peat near so much resolved as there; and the reason is plain. All the Island is covered with limestone on which the moss is spread and with which it becomes mix'd, but calcareous earth is well known to be a powerful septic, and hence it will quickly reduce the peat into an inert matter.

In short, taking all these circumstances together we are evidently led to the conclusion that all Peat or Turf consists of half putrid vegetable substances, that it is formed by the decay of woods, etc. in the same place where it is at present found.[66]

[66] Walker developed an interest in peat and soils early in life. His Hebrides report contains many remarks on soil conditions. There too he says: "All the peat mosses in Scotland which contain the trunks of trees, have formerly been woods; and from the great number and extent of these mosses, we may judge of the woods that must have existed in the country in ancient times. The formation of a peat moss, however, from a wood, requires the stagnation of water; our deep mosses are accordingly situated, either in a hollow or in a flat part of the country, where there is a subsoil that resists the descent of water; for there are many tracts of Scotland which have been formerly covered with wood, where no tree nor any peat moss appears at present. These are always dry and shelving grounds, where the water either sinks immediately, or runs off."

Walker's interest in botany led him at an early date to study and classify the plants that compose peats. Although this lecture is only an abstract of his classic paper *"On the Origin of Peat,"* most of his basic conclusions are included here. Three of these were: (1) peat is formed in place, (2) it is a peculiar product of temperate and cold climates, and (3) it is of

There is one remarkable circumstance concerning Peat earth and that is its Antiseptic power. All peat earth is found greatly to resist putrefaction. This evidently appears from the state of the wood found in it. Thus oaks are found in it as sound and fresh as when growing and capable of being work'd into furniture. We have several other instances of this. Some years ago in a Peat Bog in Ireland there were dug up some wooden bowels, some sacks filled with nuts and a coat of an unknown texture which demonstrates its antiquity.

We see the same thing in the mosses of this country from which many similar articles have been frequently taken exceedingly fresh.

I have found two instances of the Antiseptic power of peat earth on the human body. One of them is given by Dr. ———. In Lincolnshire 2 human bodies were dug out of a peat moss which were known to have been interred there 49 years before; their flesh was fair and pitted with the finger, their joints play'd freely and some new serge with which they were covered, when washed was as fresh as when it came from the Loom.

The other instance of this kind was likewise in Lincolnshire. A body which was there dug out of a peat moss had on its feet antique sandals, which plainly shew'd it to be of the Roman æra; the hair and nails were perfectly fresh and the skin when pulled stre[t]ched like doe leather.

We are acquainted with this Antiseptic property likewise in the washing of flax; flax is washed in order to putrefy the reed. But in this way when flax is put into moss water this putrefaction does not take place. And it is but of late years that our farmers have thoroughly understood, and they now carefully avoid washing their flax in water that comes from bogs or peat mosses.

Another instance of this is related by Captain Cook in his voyage round the world. When he was near the shores of Terra del Fuego he was short of water; he was obliged to water his ships with water of a brown colour something like that of Porter, and as he says exactly like that which runs from some of the bogs in England. He did this with great reluctance, and he was very cautious in the use of it, but from a little experience it was found to have no bad effect; he sailed from thence into the warmer climates, and there he declares he never had such good water aboard his ship, as it was perfectly free from any tendency to putrefaction; and hence it is probable that an impregnation of peat earth in water might in long voyages prove very serviceable to the health of seamen.

---

organic origin. A comparison of Walker's work on peat with the work of Jameson makes interesting reading.

The next article to be considered in Geology is the

SUBTERRANEAN HEAT

In general the earth is cooler than the Air in summer, and generally in the temperate climates warmer than the air in winter.[67] The existence of the Subterranean heat in winter with us is observable on many occasions, by the great degree of ——— and by the smoak of springs. This subterranean heat is more observable in sandy than clay soils, for [on] sand the Snow melts much sooner than on clay, because the sand is more porous and more pervious[68] and admits more readily the access of the subterranean heat to the snow, than the solid Clay. Snow that is lying on the ground quickly melts while that lying on a stone persists much longer because the stone prevents the access of the subterranean heat.

I before observed that the subterranean heat at a certain depth is always stationary; this was first discovered and ascertained in the Cave of the observatory at Paris by a thermometer which was placed there 100 years ago and has continued ever since, and there it is stationary the whole year round. In this cave the Thermometer remains stationary at the depth of 90 feet, but I have had occasion to observe in mines that it is stationary at 60 feet deep.

The Subterranean heat was long thought and even very generally yet considered as the same over the whole world, this however unquestionably is an error.

It was endeavoured to be proved by Mr. Mairan[69] that there was a source of heat in the earth independent of the Atmosphere and hence he accounted for the different degrees of heat in different countries.

The Count De Buffon likewise assumed this theory of his countryman; and Dr. Martine[70] was also of the same opinion. But our best Philosophical Astronomers were always of a different opinion as Newton, Huygens and Gregory;[71] they all imagined that there was no heat in the earth but what it

---

[67] In this section Walker is not referring to heat in the interior of the earth but in mines and caves at depths down to a few hundred feet. He felt that the interior of the earth was an unknown entity and that it was not possible to say anything factual about it.

[68] The distinction between porosity and perviousness was understood.

[69] J. J. Dortous de Mairan (1678–1771), French scientist, *Dissertation sur les variations du baromètre*.

[70] Unidentified.

[71] David Gregory (1661–1708), professor of astronomy, Oxford.

derived from the Atmosphere by means of the Sun; and positive experiments have since confirmed their opinion.

I may likewise here repeat that the subterranean heat of any particular place is always equal to that of the Springs, and this is likewise equal to the medium heat of the Atmosphere; by finding therefore any one of these three you have of course the other two, thus by getting accurately the heat of the Springs you have likewise that of the Atmosphere and the subterranean stationary heat.

Interior Parts *of the* Earth

We might now be led into a copious discussion of the Nature and structure of the interior parts of the earth.[72] But of its central parts we are entirely ignorant and must remain so.

The greatest discovery relative to it is the hypothesis of Sir Isaac Newton, which we have already mentioned.

By the experiments of Dr. Masqualine [at] Chechallion,[73] the density may safely be supposed to be equal to all other superficial parts of the Globe. This follows, that the matter of this globe is of greater density at the center than at the surface. Woodward and his followers have attempted to prove, that there is a huge abyss of water in the Center. Others have supposed that there is a central fire. Some appearances in nature seem to indicate that the interior parts of the Earth derive heat from a different source than the Sun and its influence on the atmosphere. At the bottom of the Cave of Observation in Paris, which is 90 feet deep, the thermometer stands all the year round at 53°, and it is always found in the other places of the globe after descending 50 or 60 feet below the surface that the temperature is always the same all the year round. Dr. Martin from this formed the opinion, that the interior heat is derived from some other source than the heat of the Sun and Atmosphere of the Earth, as it, he supposes without an immediate communication, could not, at so great depth, maintain that degree of central heat.

I am of opinion that much heat is not communicated to the central parts of the Earth by our Air; but it appears, that the regular degree of heat observed in the interior parts is derived, first from the heat of our waters sinking into the earth, and which we find in all parts of the globe, is in a medium heat betwixt the greatest heat and greatest cold of the rain that descends into the Earth in all climates.

[72] This section is taken from D.C. 2-18. Here by "interior" he means the deep interior, or core.

[73] Mount Schechallion, Scotland.

I proceed now to another subject and one of great importance, the natural history of

VOLCANOES

And here we may first notice where they occur.

On the continent of Europe we have only one vizt. Mount Vesuvius in Italy; Aetna in Sicily and some other smaller ones in that country. There is likewise Hecla in Iceland. In the Islands of the Atlantic there are several; some in the Azores, in the Canaries, and Cape de Verd Islands, some in the West Indies. There are a good many in the northern parts of the Western coast of America.[74]

In the Eastern parts of the world there are some in B[orneo] one in Sumatra, one in Japan, one in the peninsula of Kamschatka. There are some likewise in the Southern Islands of the Pacific ocean, one in New Guinea, one in the New Hebrides, and some in the Friendly Islands. There are none in the South or Northern parts of the great continent of America, [but] some lying on the western extremity, and there is but one volcano in the great continents of Europe, Asia and Africa; so that we see they are confined to a mere point in the world.

Volcanoes are generally in Islands and situated not far from the Sea, and it is plain they sometimes communicate with it, thus in the famous eruption of Mount [Vesuvius] in Italy when the Volcano was throwing out, the adjacent sea near the shore became quite dry.

Volcanoes sometimes break out in mountains, or if they break out in a plain they generally form a mountain of some considerable size.

They are every where accompanied with hot springs. Thus they are frequent in Italy which is a very volcanic country; the heat of springs is most frequently owing to their mineral contents; but Vitruvius long ago gave an observation that there were many hot springs with no mineral contents, but which deriv'd their heat merely from the hot volcanic strata they passed along. In Kamschatka there are a great number of hot springs and it is a very volcanic country; the heat of these was long ago taken by the excellent Steller and he found them to be from 74° to 200° of Fahrenheit. Some of these in Kamschatka bubble up with a considerable jet sometimes to about 5 feet high.[75]

But the most remarkable hot springs are those in Iceland which are like-

---

[74] It should be noted that Walker is referring to active volcanoes; extinct volcanoes were recognized only by the products remaining.

[75] Walker recognizes that groundwater moving along hot strata may cause hot springs to issue at the surface.

wise of different degrees of heat from 89° to 212° of Fahrenheit the point at which water boils; the most remarkable of these is the Geiser. It rises from a huge aperture of a round figure no less than 19 feet in diameter, and it throws up a jet 92 feet in height. It forms in its channel a very hard concretion[76] which is a great fossil curiosity for it is siliceous and it is the only instance of siliceous earth being dissolved in natural water.

The celebrated Bergman supposes it dissolved in the water on the principle of Papin's digester, or that the water by being subjected to strong compression became hotter than 212°. But the Gentlemen who visited it assured me that the heat on no occasion was above 212° of Fahrenheit. There are others of the hot springs in Iceland impregnated with the earth of Zeolite; there are likewise some other earthy concretions of these springs brought from Iceland that appear both different from Zeolitical and siliceous earths.

Volcanoes are most commonly preceded by earthquakes. Thus all Italy was shaken violently before the eruption of Mount [Vesuvius].

It is in consequence of this that the bursting up of a volcano frequently relieves whole countries from earthquakes, as was the case in the end of last century in the Island of Guadilupe. It was so formidably shaken for some time that it was on the point of being deserted by its inhabitants, till at length a volcano burst out in the sea in the neighbourhood, and then the earthquake entirely ceased.

Nay, even when a volcano continues to smoke and flame moderately the adjacent country is never shaken with earthquakes, but if it happens to cease then the earthquakes hardly ever fail to take place.[77]

The eruptions of Volcanoes take place much more frequently in winter than in summer. There are many appearances in Italy which give ground for this observation, and we have sufficient reason to think that it likewise takes place in all the other volcanic countries.

Besides the larger Volcanoes of Italy there are likewise a considerable number of smaller ones which they call Spiracles; these throw out both flame and smoak but in so small a degree that they cannot be perceived unless in the night time; they flame during the winter but in summer are perfectly quiescent.

It is not improbable that this more frequent eruption of Volcanoes in winter may be owing to the subterranean elastic vapours being shut up in conse-

[76] The use of "concretion" here and in the following paragraph in connection with the origin of geyserite is interesting. See also the chapter on Hydrography, n. 50.

[77] The relation of foreshocks to volcanic eruptions and the decrease in number and intensity of shocks after an eruption are now well known, but Walker's observation here is very accurate.

quence of the winter rains, because in all these countries such as Italy in the dry summer season there are cracks and chasms in the earth by which these elastic vapours escape, but in the rainy seasons these chasms being shut up the vapours are pent up.

The smoak of Volcanoes is of two different kinds the one of a Black pitchy colour; the second is of a white colour.

The first kind is evidently originated from the inflammation of bituminous matter and with this matter all Volcanoes are supplied.

But the second vizt. the white smoak is evidently owing merely to the evaporation of water, and this smoak is seldom seen unless where the volcanoe is known to be supplied at bottom with water.

It is pretty evident that all Volcanoes are considerable producers of Lightning; this Volcanic Lightning as it may be called is a well known concomitant of the eruptions of Vesuvius.

We had a remarkable instance of this Lightning a few years ago in the year 1783; in the summer of that year during the eruptions of Hecla in Iceland the Lightning continued there without intermission but in Iceland excepting at these eruptions lightning is not at all known. I formerly had occasion to take notice of that Haze that prevailed over the whole of this country in the year 1783 and during that summer while the eruptions continued in Iceland from June to September, there never was so much Lightning known in this country in the memory of any person.

One of the most remarkable Phenomena of Volcanoes is their projecting force. Sir William Hamilton[78] informs us that in the eruption of Vesuvius which happened in the year 1771 there was one stone of a very considerable size thrown out of crater to such a height that it took no less than 11 seconds in falling and if this was the case it behoved to be elevated about 1000 feet in the Atmosphere. He likewise mentions another instance which happened in the eruption of Vesuvius in the year 1779. A stone, or rather a large rock of old lava was thrown out above a quarter of a mile. This rock was 17 feet in height and 108 in circumference. We are told by Dion Cassius that in the eruption of Vesuvius which happened in the reign of Titus that the ashes from the volcano were spread to Africa, Egypt, and even Syria; this report of Dion was reckoned rather legendary but the late accounts on this subject tend to confirm it. We have one positive instance in the case of the *Resolution* one of Captain Cook's ships. When sailing within sight of Kamschatka and of a volcano there which was at that time erupting, at the time the deck of the ship was covered an inch thick with minute ashes and

---

[78] Sir William Hamilton, *Observations on Mount Vesuvius, Mount Etna and Other Volcanoes, in a Series of Lectures Addressed to the Royal Society* . . . (London, 1772).

small round stones about the size of hazel nuts, and she was then 8 leagues distant from the mountain.

Sir William Hamilton assures us (and he seems to have been at particular pains to ascertain the fact) that on one occasion there was a prodigious quantity of ashes thrown out by Vesuvius at 9 O'Clock at night and they fell at 11 O'Clock on M———— above 100 miles distant. What is the power of ordnance compared to this? Or to what cause are we to attribute it?

Volcanoes have the utmost force of fire and flame; they may explode vast quantities of the Air which is passed from a quiescent to an elastic state. There may likewise be a conflict between the melted Lava and water such as is the case between water and melted metals. But it is not even all these powers that can account for such effects.

Wherever an inflamed Volcano is supplied which in the most powerful is always the case, great quantities of steam must necessarily be formed whose expansion we know to be irresistible; the Volcano then acts on the principles of a steam engine. In this I think the great effects of Volcanoes in projecting is to be ascribed, and indeed I know of no other power in nature that can produce such effects.

The erupted matter from volcanoes is sometimes so considerable as to form hills and even mountains. Thus Mount ———— near Naples we are informed was thrown up in the year 1636 in the space of a week. The celebrated Mr. Ray view'd it about 100 years ago and he supposed it about 100 feet in height. It has since however received many considerable additions for of late years we are told that its perpendicular height is 1/4 of a mile by Sir William Hamilton, and about 3 miles in circumference at the base. This however is a rare and singular instance of a single mountain being thrown up in so short a time on a plain; yet we know that in the case of all volcanoes they are considerably enlarged and their figure altered by every eruption.

It has often happened that volcanoes have arisen in the ocean; in many places submarine volcanoes have appeared. We have one case of this in the neighbourhood of St. H———— one instance in the Mediterranean which happened in the year 1457, another in 1570 and another in 1654. On the last of these events the quantity of pumice erupted was so great that it nearly obstructed the progress of ships at sea, and it floated as far as Smyrna, Constantinople and Cyprus.

There are several instances of land being formed by them. Accordingly one of the Azores was formed in this way in the year 1628, another in that neighbourhood in 1638, and one in 1720 about 17 Leagues southeast of the Island of ————.

It is very probable that the 7 Eolian Islands in the Mediterranean have been formed in this way as they consist entirely of volcanic matter. They were visited by Kircher in 1638, but they were then altogether quiet and no fire in any of them but one.[79]

In 1683 we had an account of an Island having been thrown up in the neighbourhood of Iceland. The accounts given of it were very short and seemingly authentic and several ships in these seas met with great quantities of Pumice, but the next year, when a ship was sent out by the King of Dane- mark on purpose to collect information concerning it, it could not be found so that if the report was true concerning it, it must have been again im- mersed by another shock.

Besides the present volcanoes it is likewise evident that there have been former ones which are now extinct; nothing can be more clear than that they have subsisted in Madeira. The same is the case with the Peak of Teneriffe; several have likewise existed in the Moluccas, in the Andes, in Italy and in the West Indies; but then the traces of Volcanoes are there very plain. You find in all of these places pure unconnected Lava and likewise great quanti- ties of Scoria. You find also ——— that peculiar products of Volcanoes and likewise Pumice; and indeed it is but fanciful to imagine a volcano has ex- isted in any place unless you there find its peculiar products.[80]

The Island of Ascension was certainly in former times a volcano. The whole of it almost entirely consists of Volcanic products; it would seem too that it had had eruptions at no very remote period, for tho' of considerable extent yet when visited about 40 years ago by Osbeck,[81] the pupil of Linnæus in his way to China, he found on it only 3 species of plants. All the [other] vegetables had certainly been extirpated by the Volcano, and they were either the only plants that had escaped or the Island had acquired them in the course of time by seeds from the ocean. They were likewise in very small quantity, for of the *Sherardia Fruticosa* which was one of them he only found one plant and certainly if he took it up he was a hard botanist indeed.

But of late years the Island of Ascension has been more accurately ex- amined, and in the more remote parts several natural rocks of Limestone have been discovered and on these several other species of plants have been found. There have likewise been found in that Island branches and trunks

---

[79] Walker was fully aware of the existence of extinct volcanoes, identifying them by the presence of old volcanic matter. His remarks on submarine volcanoes, the production of new islands by vulcanism and the subsequent disappearance of such islands are interesting. Also interesting in this passage is his correct identification of several places as volcanic in origin.

[80] Walker did not recognize the volcanic origin of rocks of the Edinburgh district.

[81] Peter Osbeck, *A Voyage to China and the East Indies* (London, 1771).

of trees petrified which shews that it has not been thrown up from the sea, but after the Island had been formed a volcano had taken place.

I come next to notice the

Peculiar Products *of* Volcanoes

The first of these is Lava. It is common to denominate Lava everything thrown out by a volcano but the term is properly to be applied to nothing but only completely vitrified matter and accordingly among the exuviae of all volcanoes are found great quantities of this glass. But the greatest proportion of matter thrown out by Volcanoes consists of Scoria or half vitrified matter, but indeed this is commonly termed Lava, though improperly.

There is another product of volcanoes, vizt. capillary Glass. The matter of the Volcano in this case is so completely vitrified, that by the mixture with water it is separated, and formed into small filaments.[82]

A few years ago there was a shower of such glass filaments from the volcano in the Island of Bourbons to the distance of six Leagues, and in Ireland during the eruption of Hecla there was such a great quantity of those filaments that they covered the whole face of the country to an inch in depth, and when a light wind took place they were blown about like small pieces of wood. It is precisely the same with what takes place in an Iron furnace by a great heat and a strong blast: part of the matter is detached.

There is another product of Volcanoes which was first noticed by Cronstedt and I do not find it mentioned by any other author; the *pearl-slag*.[83] I suppose he had it in Sweden for I never heard of it being found in Italy, but I have lately had sent me some fine specimens of it from Madeira. It consists of small round nodules of the size of peas and is chiefly composed of Calcareous earth.

The next of the volcanic products is the volcanic ash. This is thrown out in prodigious quantity. It is of a loose pulverulent appearance mix'd with small stones. It was with a deluge of this that the cities of Herculaneum, Stabia and Pompeia were overthrown.

What is commonly called Tupha is likewise a peculiar volcanic product, but then it seems nothing else but the volcanic ashes consolidated and become Lapideous in the course of time. And accordingly in the neighbourhood of Vesuvius they have it in all its various degrees of induration up to the state of stones fit for buildings, and accordingly most of their buildings are formed of it.

[82] Pele's hair.

[83] The name pearl slag is introduced here from another manuscript, D.C. 2-24.

There are likewise found several other smaller and more curious products of Volcanoes such as those called the Lava garnets, the Lava Gems etc., and besides these a number of other fossils are generally found in the craters of of Volcanoes, as salts, such as *Sal Ammoniac*.

These are the prevalent products.[84]

We now proceed from the consideration of volcanoes to that of

EARTHQUAKES

And before considering their cause, it is proper to narrate their Phenomena.[85] These may be divided into three heads, vizt. the previous, Concomitant and Consequential Phenomena of Earthquakes.

Great alterations in the mineral kingdom have been ascribed to these. Among these Phenomena we may remark, and it will readily occur to every person, that Earthquakes are most frequent near Volcanoes and hot Baths.

This is the case in the Islands of Kamschatka, Japan, and Iceland. Earthquakes seem chiefly to affect Islands and maritime places more than Inlands and Continents. None remarkable have happened on any great Continents or in Inland Situations; but all remarkable and considerable Earthquakes have happened in Islands and in Countries near sea shores. What I observed concerning Volcanoes seems true concerning Earthquakes, that they are more frequent in winter than in summer, as is proved by long and accurate observations.

The island of Jamaica has been tormented with them often, and there Earthquakes have produced most dreadful effects. They have often had

[84] Among the products recognized by Walker are lava, basalt, scoria, cinders, glass filaments (Pele's hair), pea-sized fragments, ash, tufa, pumice, salts, and gases. He recognizes that volcanoes repeatedly erupt and that their cones grow larger with time. He gives us no unusual explanation of the origin of volcanoes but describes them in some detail and identifies their products. Basalt (basaltes) is mentioned at various points in the text. Walker includes basalt as a volcanic product, and he considers whinrock a kind of basalt, although he is confused about its origin. The volcanic formations now known as dikes and sills baffled him. He knew that the whin at Salisbury Cragg was basalt but did not think it was volcanic. The dark smoke from volcanoes he falsely attributes to bitumen. And finally although he appreciates the "extrusive" processes, he has no clearly defined concept of "intrusive."

[85] The entire section on earthquakes is from D.C. 2-18 and a similar section may be found in D.C. 2-24. This section is not a particularly strong one. Although Walker recognizes the relationship between volcanic eruptions and earthquakes, he says, "I can offer you nothing satisfactory even to myself," on the cause of the second class of earthquakes. His guess is that they may have some relationship to inflammable material. Walker's knowledge of earthquakes is normal for the time. Perhaps his only significant contribution is the observation that structures built on solid rock withstand shocks better than those on alluvium. His comments on the vibratory nature of the shock and his reference to aftershocks are interesting, as is his remark that major earthquakes recur at the same point over a period of time.

terrible ones about the end of last Century and beginning of this. Earthquakes always happened during the rainy season almost annually. In summer the ground of Jamaica chaps considerably from heat, leaving cracks and fissures from 1 1/2 feet to no less than 2 feet wide in many places, which are of considerable depth; and then the people dare not ride upon the fields, but must keep the made roads, and while the fissures and cracks remain open no Earthquake happens in the Island. But when the rainy season comes on, which is their only winter, then these Cracks and fissures are closed and shut up; at which time the Earthquakes commence.

Observations of the same kind were made by Dr. Shaw[86] who resided in Algiers 12 years Chaplain to the British factory there, where they always happened after hard rains had closed the fissures of the ground. This leads us to conclude, that the inflammable air or Sulphurous vapours are pent up and not suffered to escape without the heavings of the Earth.

As to the

Previous Phænomena *of* Earthquakes

From all accounts we have of Earthquakes, they are preceded by calm hot air and weather with a lurid appearance of the atmosphere; hollow rumbling noise in the Earth like of a Cart or Waggon on a rugged Causeway. The Barometer is always found remarkable low, and [is] affected from great distances from the seat of the Earthquake. For the late Earthquakes at Calabria 4 years ago affected all the Barometers in the North of Europe.

Next we shall mention the

Concomitant Phænomena *of* Earthquakes

One of the most remarkable is the violent motion very perceptible of the waters; if the earthquake happens at a maritime place, the sea first retreats and immediately returns with dreadful violence. This was remarkably the case in the great Earthquake which happened at Jamaica.

At Lisbon, when the dreadful Earthquake happened there, all the waters were greatly agitated, and those in wells and pots were thrown out at the top with such violence that they ascended some fathoms in the open Air.

In the Cisterns of Nevis and Antigua the water is often urged out at the top when the Earthquake is at hand. A Gentleman of my acquaintance, who resided at Lisbon, and who had observed many shocks of Earthquakes, contrived a machine by which he could measure the violence of the shocks in earthquakes by means of bowl of water lined with chalk.

[86] George Shaw (1751–1813), assistant keeper of natural history in the British Museum.

The most remarkable earthquake that ever happened in Britain, which we have any account of was the 1st November, 1755, at the same time that the great Earthquake happened at Lisbon. A most sensible commotion of all the standing waters was observed, and they were thereby greatly affected.

In Lisbon all the Buildings on sand and gravel were brought to the ground, but those built on rocks withstood the shock. In the Earthquake at Jamaica the same thing happened at Port Royal; all the Buildings which had rocky foundations withstood the shocks. And in the Island of Sicily, where Earthquakes frequently happen, the same thing happens; but all houses of Stone and Brick in every place if the shock is violent, must and actually do tumble down to the ground, while those of wood remain entire. The motion, produced by Earthquakes in the Earth, seems vibratory; and by the sudden jerks the stones and bricks are dislocated at the joints; whereas buildings of wood are strongest and compactest at the joints and stand the shock of course without injury.

An eruption of sulphureous waters having noxious steams and exhalations frequently accompany earthquakes.

Thus the hot Baths in Italy are often increased in their quantity of waters by earthquakes.

Another effect of Earthquakes is, that running water is frequently absorbed by them. Sometimes the fresh water instantly becomes salt, especially those near the sea. Thus the pools and lakes of fresh waters in Jamaica were turned salt instantly.

The last great and tremendous step, and I may say the greatest effort of Earthquakes is the bursting asunder the surface of the earth, which only happens when the Earthquake is of volcanic nature. They are usually accompanied with a volcanic eruption. There are instantly formed great chasms in the Earth; from these chasms burst forth surphureous vapours, fire, air, flames and hot boiling waters.

On some other occasions the surface of the Earth subsides; the Earthquake sinking the ground. Thus in Jamaica on the side of a mountain a piece of land, three leagues long sunk and was converted into a lake of water of considerable depth by the efforts of the Earthquake.

And as to the phenomena that usually follow Earthquakes, or the

Consequential Phænomena *of* Earthquakes

The first of these; that every considerable Earthquake is usually followed by a number of small ones. And to these places, where Earthquakes have been, they generally return again to the same place. This is remarkable at

Lisbon, as it has been often visited in 3 or 4 Centuries, when none such happened in any part of the coast of Portugal, nor any where in France or Spain.

In 1653, in the great Earthquake at Sicily, no less than 90,000 people perished. This Earthquake affected all the same tract of Country, and was as it were confined to one chain of mountains, towns, and Castles, as it had done in a former Earthquake, which happened in 1643.

Earthquakes are generally followed by epidemick sickness, either of a local or general nature. And finally, the annoyance of the Earthquakes to health, is remarkable by the breaking out of a volcano, as at Guadaloupe. Volcanic Eruptions, sometimes following, also relieve the Country from Earthquakes as I have before noticed.

And as to the

CAUSE *of* EARTHQUAKES

I can offer you nothing satisfactory even to myself. It is needless to notice or examine the many idle opinions given of this cause by others. I only say in general that Earthquakes are of two kinds.

1st. *Those of a Volcanic Nature*

The causes concurring here are inflammable matter in the Earth and the earth of a cavernous structure. This inflammable matter, being from some cause reduced to an ignited state, produces boiling water. The steam of which being condensed, and that having an almost irresistible force, bursts and breaks the Earth and produces those hollow rumbling noises which are then heard.

But as to the

2nd Kind of Earthquakes

This Class seems essentially different, vizt. those in which there are certain tremors over particular tracts of the earth, and are not volcanic in their nature, and which are never accompanied with any eruption at the surface. Such are the number of small Earthquakes which happen in England, and on the neighbouring Continent. Some have ascribed the causes of these to the effervescences of acid and alkali in the bowels of the earth. Some have ascribed it to a certain etherial matter. Others to the expansion of confined air. And some again to electrical fire.

In Scotland there are few if any Earthquakes, and those that have been observed are of no importance. Therefore we cannot expect Volcanoes here, any more than we can expect a Volcanic Earthquake; and for this reason, that we have no volcanic or . . . inflammable materials for breeding or feeding Volcanoes.

In 1700 the most remarkable Earthquake that we have any account of happened in Scotland. It was strongly felt from one extremity of Fife to the other, even from the Promontory of Fifeness to Dunfermline. But at this time the Coal works of Dysart in Fife were then on fire, which raged with great fierceness; and I imagine, that the appearance of this Earthquake was from that cause.

We have now gone through a great variety of curious and entertaining particulars in the natural history of the Earth; there is but one now remains, which seems to have been unaccountably neglected by Physicians; that is

The NATURE *and*
QUALITIES *of* SITUATION

I shall make a few remarks, which is all I am able to do on the subject.[87] It is plain, that the wholesomness or unwholesomness of a place depends on the qualities of the Air. It appears that there is little difference with respect to the rarefied or dense Air. In the varieties of these, there are many level champaign countries, little elevated above the level of the Ocean, whe[re] the air consequently is very dense, in which people enjoy excellent health. On the other hand, in the Vale of Quito, 9,000 feet higher, the Air, though proportionally rare, is exceeding wholesome, though the contrary has been asserted by Travellers, who said that those who ascended high mountains were taken with spitting and vomiting of blood, etc., for the French academicians lived several weeks upon a mountain 9,000 feet high, and felt no greater inconvenience than cold. The unhealthfulness of Air is unquestionably owing to its impregnations. It is impregnated with mineral, vegetable, and animal substances. The fossils that impregnate the atmosphere, are few, vizt. Aerial acid, sulphur, Arsenic, and inflammable Air, and their quality, when volatilized, is comparatively small, but the first is most so dispersed through the Air as to be harmless, except in the Grotto del Canii. The two next are only known in Volcanic Countries. The last by its levity rises far out of reach.

As to the impregnation of Air from Vegetables, it is to be divided into two kinds.

1st. However agreeable the odour of some flowers may be and some fruits, it is evident they transmit an air highly phlogisticated. All vegetable putre-

[87] This section is from D.C. 2-18.

faction is likewise noxious. But there is another kind [of] impregnation from Vegetables, which exude a dephlogisticated Air that is pure and salutary to the highest degree.

Animal impregnations are always unwholesome. The breath of animals is considered as such, and so is the effluvia of all putrid animal substances, the vehicles of contagion. Such animal impregnations are to be dreaded; more stagnation increases their virulence, free ventilation abates its noxious powers.

But the most frequent impregnation, and what is more or less unwholesome, is from water. Dry Air is always the wholesomest, but the impregnation of Air from stagnated water, is undoubtedly very pernicious. It is now well known, that phlogisticated air is improved and rendered salubrious by agitation with water, and on this perhaps depends the wholesomness of sea Air.

I know but one exception to the unwholesomness of stagnant water, and that is the stagnant of [peat?] Bogs, as in Ireland. The people who live near them are quite free from diseases. It is on account of stagnant water, that all low situations are less wholesome than those that are high. It is likewise pretty certain that there are many acute diseases that are more virulent in some climates than in others; and they are generally found most virulent in low situations, while the higher are perfectly free from them.

I have observed an epidemic fever every year in the low parts of Anandale, but it is never known to touch on the very high grounds or dry situations. Certainly there are many acute diseases, especially the epidemic diseases of Children, far more virulent, and much more dreadful in some places than in others. A rich soil is accompanied with the most unhealthy Air; as happens in the southern Shires of England; and where the primary mountains, soil and strata exist, there we will always find the most healthy situation. And on the contrary, in countries where [there] is an extent of secondary Strata, especially of earthy matters, there the atmosphere is in the most unhealthy state, and consequently the air of any Country or Tract is most salutary, where there are many pure and excellent perennial springs; and where the ditches are dry, it may be judged healthy. But where the ditches of a Country are wet, I immediately conclude, that the situation of that Country is sickly and very unfavourable to human health.

These hints I thought it was proper to throw out; they can serve no other purpose, than to induce others to enquire into these topics to a greater length.

Thus, Gentlemen, we have finished the three first divisions of Natural history viz., Meteorology, Hydrography and Geology. We shall next proceed to consider the three last which are denominated the

IMPERIUM NATURÆ

OR

EMPIRE OF NATURE

VIZT.

The Fossil, Vegetable, and Animal Kingdoms

But before I proceed I will make some general Observations on the three kingdoms of Nature in general. (See Soam Jenyns,[1] his first description on the Chain of Universal Being.)

These form the great Scale of Nature. To comprehend a complete knowledge of which Scale every climate, every part of the Earth, and all its productions and every thing in it is required to be examined. At one end of this Scale is placed dead, inert matter, and at the other end of the scale is situated Man. In consequence of the imperfection of our knowledge, this chain of our being must be very imperfect.

This short chapter is from D.C. 2-18, where it follows the section printed at the end of the last chapter.

[1] Soame Jenyns (1704–87), English politician and writer.

The mineral kingdom affords but a very few links of this remarkable chain.

From inert and dead matter the Scale proceeds through the mineral kingdom where we observe something, which in some degree approaches to, and has the appearance of organized matter, and there is nothing that comes so near it as the crystallization of salts, and notwithstanding all that has been said on the subject, it can never be accounted for on mechanical principles. There seems to be something (a *vis insita* if I may call it) that disposes them to assume those different forms, independent of any mechanical law whatever. But though the crystallization of Salts has somewhat of the appearance of organic matter, yet the links of the Chain, of which, it must be observed, the mineral kingdom affords but few, the passage betwixt the mineral and vegetable kingdoms is abrupt; because it is a passage and transition from dead to living matter. But the passage betwixt the vegetable and animal is not so. As on the other hand, the vegetable blends intimately with the animal kingdom running, like the shades of a painting, imperceptible into one another; so that with the most scrupulous nicety and most accurate observation we cannot say where the one begins and the other ends. This was the case with ancients; for, what they thought the confines of these two kingdoms were easily distinguished and well marked; but by the great discoveries made by the modern Philosophers on the Lithophyta, Zoophyta, Corals, etc., it now seems very difficult to ascertain these limits. I would make a remark here. Where the animal kingdom begins, the progress of life is more obvious than in the vegetable and more striking through the whole of the kingdoms.

There we find an infinite variety. There we find the great ends of animal life answered by an infinity of beings. It must however be acknowledged, that in this state there are many chasms; but we behold such a concatenation as persuades us, that all beings are arranged in a progressive order. As for the smaller varieties, they are no more than the parts of the division between one being and another. Every void we perceive is probably owing to the imperfection of our knowledge, to our inability to investigate them; and we know that this scale is gradually becoming more and more compleat as the degrees of our knowledge become more extensive.[2]

Each species of being, whether fossil, vegetable or animal, approaches to some other; and the same thing also reaches from these others in something else, and thereby is readily distinguished by some characteristic from others.

[2] The great gaps and voids in the chain of life of which Walker speaks have for the most part been filled since his time by discoveries made in paleontology. It is to Walker's credit that he recognizes the gaps and believes that such voids will be filled when knowledge is more complete.

Maupertuis,[3] struck with the great voids he finds in nature, proposed an hypothesis: That the Earth had once a conflict with a Comet, and fell into confusion from being affected by the Comet, which destroyed many species of Vegetables and Animals, and produced Chasms and void spaces in the chain of Beings. But this I would treat as mere theory, and I do say that such an opinion as this is unworthy of a man who had any name in Philosophy. Were we to suppose this, the meanest of mankind might, on such chimeras, account for anything. But it is unphilosophical to reason in nature except from what we know.

When we come towards the head of this great Scale of Nature we meet with one thing that has puzzled the greatest Philosophers to account for; and that is "where instinct ends and reason begins."

One of the great so remarkable spaces in the great chain of being is that subsisting betwixt man and the inferior animals; that is, the limits betwixt the instinct of the brute, and the reason of man. We observe in man the powers of reason and intellectual abilities greatly diversified in different individuals. We observe the same in the inferior animal with respect to instinct; as some beings are superior to others in the same species. This is finely expressed by our moral Poet.

> How different Instinct in the grovelling swine,
> Compared, half reasoning Elephant, with thine!
> 'Twixt that and reason what a nice barrier'!
> For ever separate, yet for ever near.[4]
>
> POPE

It is unquestionable, that the intellectual powers of man are clearly superior to those of the inferior animals—that he distinguishes clearly and certainly a moral sense by a power of conscience and a sense of religion. There is likewise another remarkable instance of distinction that subsists betwixt them, which I call Specifical Improvement. For example; There are some Animals of high capacity of penetration, but this they hold only while they live, and then extinguished they cannot communicate it to their posterity or co-temporals. But however different this is in mankind, from him we have a Specific improvement. The wisdom and improvements and acquirements of one man being communicated to his fellow creatures and his discoveries are perpetuated from one age to another.

But the most distinguishing mark between man and other animals, is the principle of religion. Of this they are totally defective. And surely it never

[3] Pierre-Louis Maupertuis (1698–1759), French astronomer.

[4] Alexander Pope (1688–1744), "Essay on Man," Epistle 1, Section 7.

be sufficiently lamented, that this principle should ever be debased, as it is the only principle that serves to distinguish us from the brutes, and the pride and glory of man.

Now as we are about to take a view of the three kingdoms of Nature, it is proper to make some remarks on the nature of organized and unorganized bodies.

By organized bodies is meant the whole body of the vegetable and animal matter. By unorganized Bodies, is meant the fossils. Were we to give an exact definition of Organization, it would be this: "It is the motion of organic fluids in organized bodies."

Fossils have no containing vessels or contained fluids; they are of course destitute of Organization and of life. It is true in some bodies we meet with something that approaches to organization; particularly in the crystallization of Salts; which seems to have something of Organization; but the fact, however, cannot be admitted; that it does not depend entirely on mechanical principles, I have already shown.

We are next to take a general view of the number of Species: in each of these kingdoms of Nature, and in them altogether what they are supposed to consist of.

In the

Fossil Kingdom

we have an accurate system delinated by Linnæus, but so extremely limited and circumscribed, that it is rather to be considered as relating to Sweden alone, and a history of the fossils of that kingdom in particular, than of those of the globe in general.

Dr. Woodward was the first, who formed a great System of Fossils, of which he published catalogues. In his work we find mentioned and described in some degree 3314 species and varieties of fossils. The Authors who have added more since Woodwards time are Wallerius[5] of Sweden, Dr. Hill,[6] Mr. Forrester,[7] Dacosta.[8] From the additions made by them we may assume, that there are discovered of Species and varieties of fossils better than 5000.

[5] J. G. Wallerius, *Mineralogia, eller Mineralriket indelt och Beskrifvet* (8 vols.; Stockholm, 1747).

[6] John Hill, *Fossils Arranged According to Their Obvious Characters; with Their History and Description* . . . (London, 1771).

[7] Perhaps Johann Reinhold Forster (1729–98), *Observations Made during a Voyage round the World* (1778), or his son Georg Forster (1754–94), *A Voyage around the World* (1777); both men accompanied Captain Cook on the 1772 voyage.

[8] E. Mendes da Costa, *Natural History of Fossils.*

Next as to

NUMBER *of* PLANTS

The whole number of plants mentioned and described by the ancient Naturalists Aristotle, Dioscorides, and Galen, Theophrastus and Pliny, do not amount to above 800 Species.

In modern times, the first great collection of plants made in any work was that exhibited by Caspar Bauhine[9] in his *Pinax;* in which 6000 are fully enumerated.

The first universal catalogue of plants which we know is that of [Tournefort][10] in his *Institutions.* They contain betwixt 8 and 9 thousand. After this publication, he traveled to the Levant and in his *Corollarium* he adds 1300 more. So that in his day 10,202 Species of plants were known. Soon after the publication of Tournefort's *Institutions* Dr. Shuzer [Scheuchzer] was at great pains to preserve the number of plants that were then known; and he, in 1719, raised the number to 16,000; and from that time we have no general history of plants till the year 1753. Then Linnæus published his first Edition of his *Species Plantarum;* wherein 7500 species of plants, exclusive of the varieties, were delineated. And this was also exclusive of the varieties he had detected in the Catalogues of Bauhine, Tournefort etc. Though Linnæus delineated 7500 Species, yet in his preface he talks of a number that might subsist, that he has not observed; and was of opinion, there was on the whole globe not above 10,000 Species. But in my Opinion, I think, there is a probability that he is much mistaken in his computation, when we reflect on some of the late discoveries in Botany, as Rhumphinos[11] history of Amboyna, that is, the plants of the place; where we find above 1000 Species of Plants altogether unknown to former Botanists. And when any new Country has been visited, there has always been found plants not before known. So that the number of plants is still increasing in this way; as happened to Sir Hans Sloan when he visited Jamaica. In that Island he found 800 new Species. Now Dr. Brown[12] has discovered 1200 Species more in the same place, which Sir Hans Sloan never beheld. Dr. Wright[13] has still added many hundred species more to both, which were never before known. Therefore I think

---

[9] Kaspar Bauhinus, *Pinax theatri botanici sive index in Theophrasti, Dioscoridis, Plinii et botanicorum qui a sæculo scripserunt, opera . . .* (Basel, 1623).

[10] Joseph Pitton de Tournefort (1656–1708), professor of botany at the Jardin du Roi.

[11] G. E. Rumphius, *Herbarii amboinensis auctuarium* (Amsterdam, 1755).

[12] Patrick Browne, *Civil and Natural History of Jamaica* (London, 1756).

[13] William Wright, *An Account of the Medicinal Plants Growing in Jamaica* (London, 1787).

that I can show you, that this Island of Jamaica alone produces betwixt 2 and 3000 plants.

Circumnavigators report and have shown us, that the plants of the southern hemisphere are different from those of the north. Nay even in the northern hemisphere we know but little of the plants in the interior of the Continents and but a few places on its shores. Putting these and other observations together, and surveying the whole, I am of opinion, there subsists upon the surface of the globe no less than 20,000 species of plants at least.[14]

And as to

ANIMALS

We never had any considerable catalogue of them by any single writer till Linnæus published his *Systema Naturæ* in 1766, and delineated 5960 species of animals in this work. But there is reason to believe, that there have been detected by Naturalists and described, more than 10,000 species in the animal kingdom. And there is reason to believe that their number is little inferior to that of plants; and in my opinion they may amount to 20 or 25,000 Species. And then in all there exists on the globe about 70,000 species and varieties of animals, vegetables and fossils, that can and ought to be clearly distinguished from one another. From this immense number, therefore, you must needs be sensible of the necessity for method and arrangement in the history of Nature; for method in all things is necessary where the parts are many in proportion to the number of those of which the subject is composed. And as natural history comprehends all the species and varieties in the creation, we can not arrive at a knowledge of them without this arrangement and method. But the prosecution of that plan is the great work of natural history laid down by Lord Verulam, the great restorer of modern learning; who saw two Centuries ago, and even deplored the want of a proper method. He projected the ground work, the accomplishment of which has gradually been carried on by all the following Naturalists, and now is brought to a considerable height. There are indeed some Philosophers of this Age that undervalue the usefulness of method and arrangement in natural history; as Buffon. Gravellin[15] talking of him says "But it is too plain, that when Buffon came on the subject, his rancor against Linnæus is too visible." He is followed by Mons. Bounet,[16] who vilifys method and terms it mere nomenclature.

[14] George Neville Jones, "On the Number of Species of Plants," *Sci. Monthly*, May, 1951. By 1814 the number of plants listed was 32,000.

[15] Unidentified.   [16] Unidentified.

But on that account we see great defects in his works; for when he describes, we know what he so describes exists in nature, but how to find the object, we are ignorant. Buffon in his work reduces animals under two general heads only; the Domestic and Wild Animals. But he does not know names himself; at least he affects to despise; and therefore we find him quite confused in his work.

There is no Science that stands more in need of arrangement than natural history; without this it must remain, *"rudis indegestaque moles."*

We have great reason to believe that the different number of species remains the same as at the creation;[17] and how they have been protected is exceedingly curious; as amongst the animals we meet with many that are devourers of others. But these are necessary as a barrier. Nature has prescribed boundaries to all; and, to all these, a law seems to be laid down: "Hitherto shalt thou come and no farther."

METHOD *in* NATURAL HISTORY

is reducible to two points: The one is Definition, and the other Division.

In the arrangement of the bodies of natural history the Aristotelian rules of division, by Genera and Species, must be followed. They are found to be the best of any of human invention, and are sufficient to answer every purpose in modern times. By these rules the different modes of arrangement have been formed by the Linnæan method, appear to be the best; and therefore have gained over all the nations of the earth an universal approbation.

He takes five divisions or members:

> Classes
> Orders
> Genera
> Species and
> Varieties.

The members of this division are entirely arbitrary; but at the same time it is most commodious. It is here as in true Logic. We can only arrive at rules of just arrangement from the similarity of Species to constitute Genera; from

[17] After a full and accurate review of the number of plants and animals described and after predicting that many more species would be found in various parts of the world, Walker comes close to the whole truth but then slips when he states that no new species have been added since the creation. Although he believed, strongly, that all life was gradational and that no gaps existed in the scale of life from the lowest form to man he did not comprehend the vast number of species of fossils or the dying out of old species and the introduction of new.

the congruity of Genera to constitute Orders; from the affinity of Orders to form Classes. Next, the Characters of the members of these divisions are properly considered by Linnæus in a threefold view.

The Characters are
1st Naturalis
2d Essentialis
3d Factitious

First as to the

NATURAL CHARACTER

It is a delineation and description of the whole parts of the body. Boerhaave in his work on plants does not deliver a few artificial characters of a Genus, but gives a lengthened character, containing a just accurate and important description of all the parts. This idea was seized upon by Linnæus who from him forms his natural genera of Plants in his *Genera Plantarum,* which contains a just and scientific description of all the parts of the fructification of Plants.

As to the second, vizt. the

ESSENTIAL CHARACTER

This differs from the natural character, as consisting only of one individual character, by which a particular body is clearly distinguished from all others. This sometimes occurs in the Species, but seldom in the genera; and still rarely in the division of Orders and Classes.

Thus the Nectarium of Flowers is found in those, the base of which possesses a Nectarium, as in the *Ranunculus* etc. Thus in the whole Genus *Ranunculus,* the Nectarium is found at the base of the Petal, and is therefore very properly made the essential character.

The 3d sort is the

FACTITIOUS CHARACTER

This must take place in an arrangement, where a species or genus has the essential or general Character wanting. It consists of several parts or subordinate characters; every one of which is possessed by another genus; but no genus is possessed of the whole. The whole or most of Linnæus' Characters are of this kind.

In respect of method too, there is another triple division, vizt. as

Natural ⎫
Artificial ⎬ Method
Mixed ⎭

The Natural Method

Is this: It consists not in many single characters, but takes in the whole characters *in cumulo;* and to these joining general habit and other properties, its place therefore is not to be derived from a particular part, but from the whole taken together.

The Artificial Method

Consists of this: When there is one fix't character established, and all the bodies that possess this character are reduced under that division. There are many instances of this among Naturalists; but none is strikingly artificial as the vegetable system of Dr. Hill.

The 3d and last is the

Mixed Method

It is of that sort which participates both of the natural and artificial parts, making the one give way to the other occasionally. The most of the methodical arrangement of the three kingdoms is of this kind, as in Linnæus' *Species Plantarum;* and in it he says, that *Salix* has but two Stamina, whereas some have 3 or 4 but they are in other respects genuine *Salices*.

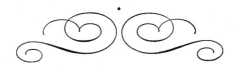

I shall now proceed to particulars and begin with

MINERALOGY

Previous to our entering more particularly into Mineralogy it will be proper to take a general view of its Literary History.

The first writers that we have on the subject are Aristotle and Theophrastus among the Greeks. In their writings however we find nothing like arrangement other than one general division which they made of the fossil kingdom; they divided all fossils into what they called *Feculia* and ———.

By the first of these Aristotle understood all those fossils immediately and entirely earthy.

By the second he meant all the metallic bodies and such as flow either through fire or water.

The next writer on Mineralogy is Galen;[1] he divided all fossils into earths, stones and metals.

Pliny is the only other author among the Ancients who wrote on this subject. He delivers a very copious account of fossils in 4 books of his natural history but without any arrangement.

In short what we have left by the ancients on the subject of Mineralogy is next to nothing.

The first writer on this subject after the restoration of Letters was the

The material on mineralogy in one of the manuscripts immediately follows the "Imperium Naturae" material. Hence the lead-in sentence at the top of the page.

[1] Galen, *De simplicium medicamentarum.*

famous Conrad Gesner.[2] He was the first person of considerable note who attended to the subject and attempted to distinguish fossils by Genus and Species.

Next to him there was a very remarkable man of that Æra the celebrated Cardan[3] the most surprizing man and probably the most learned that ever appeared, his works shew every where the most profound erudition, and he was known by the name of *"Magister centum artum."* This extraordinary man came to this country before the Reformation. He had then the greatest repute as a Physician, and he came from Pisa to cure the Arch-Bishop of St. Andrews; he collected several fossils in Scotland and some I find in the Kings Park here. In his work he is very copious on the subject of fossils, but seems totally unacquainted with method or arrangement.

Next to him appeared Georgius Agricola[4] the German. He divides all fossils into earths, stones, metals and a peculiar class to which he first gave the name of *Lucci concreli,* under this title comprehending all the Bitumens, the Sulphurs and Salts. He had the merit of being the first who distinguished between the simple and compound fossils, and was the first who attempted to distinguish the different earths not according to their uses as was formerly the case but from their external qualities or sensible properties.

About this time there also appeared a remarkable man in Mineralogy, Fallopius.[5] He was the first who treated on the subject of Mineral waters. He divided the fossil kingdom into earths, stones and Metals. There is one thing to be observed in the writings of this old author. In modern times there was a great discovery struck out by Mr. Pott[6] of Berlin; it was that all the primitive earths by themselves are apyrous, but this discovery which has always been attributed to Mr. Pott is clearly pointed out in the writings of Fallopius.

Soon after this there appeared Cesius.[7] He published a very voluminous history of fossils, but it is only a mere compilation from the works of former authors.

The first persons among the modern that aimed at the proper method of

[2] Conrad Gesner, *De rerum fossilium.* . . .

[3] Girolamo Cardano, *Several comentarii.* . . .

[4] See Georgius Agricola, *De re metallica,* translated by Herbert and Lou Henry Hoover (New York, 1912).

[5] D. Fallopius, *De medicatis* . . . (1564).

[6] J. Pott, *Lithogéognosie* . . . (1753).

[7] Bernardius Caesius, *Mineralogia* . . . (1636).

arrangement in the fossil kingdom were Becher[8] and Stahl.[9] They were followed by Henckel[10] who first endeavoured to divide them into Classes according to the earths of which they are composed.

Then appeared Newman[11] who attempted to detect the principles of fossil bodies by the means of Acid Menstrua. His experiments were all in the humid way.

But soon after him appeared the celebrated Mr. Pott of Berlin, who in 2 small octavo volumes entitled *Lithogeognosia* has enumerated no less than 2000 different experiments all made in the furnace and he was unquestionably the first who by means of fire laid open the constituent parts of fossils. He found that all the fossils were composed of 4 primitive earths. He fix'd these to be the Argillaceous, the Calcareous, the Gypseous, and the Chrystalline or Vitrescent. It is surprising however that Mr. Pott should have considered at least two of these as primitive earths. For by the Argillaceous earth he means not pure Argilla or the earth of Alum, but common natural Clay, and that we know is compounded of the Earth of Alum and Siliceous earth; the other is the Gypseous earth which is by no means a simple earth being compounded of Calcareous earth and Vitriolic Acid.

Mr. Pott was followed by a German author Woltersdorf[12] who endeavoured to erect a system on these 4 primitive earths which Mr. Pott had discovered.

Linnæus next published his *Systema Naturæ* the third Volume of which contains his Mineralogy. Notwithstanding all its defects it must be allowed to be the first real Systematic work on Mineralogy, and notwithstanding it is erroneous in many things in point of principle. Yet there is in it an accuracy and neatness, especially with regard to the specific distinctions which gives it a degree of utility.

After Linnæus there appeared in Sweden too a very remarkable man in Mineralogy, Cronstedt a Swedish Nobleman,[13] a man undoubtedly of deep research and great genius. He published what is called an essay on the Mineral System; as an essay it is a work of prodigious merit. It was first published in the Swedish language, then in the German and from the German

[8] J. J. Becher, *Mineralogia* . . . (1662).

[9] G. E. Stahl (1660–1734), primary author of the phlogiston theory.

[10] M. J. F. Henckel, *Pyritologie* . . .(1725).

[11] Kaspar Newman (1683–1737), professor of chemistry, Berlin.

[12] J. L. Woltersdorf, *Systema minerale* (1738).

[13] Axel von Cronstedt, *Försök til Mineralogia eller Mineralrikets Upställning* (Stockholm, 1758).

it was translated into English by Engerstroom.[14] The great merit of this work is that he fix'd by experiments, clear and distinct Characters of the Classes or Orders of Fossils, for properly speaking it contains no generic or specific distinctions.

Since that time it has been lately republished by Mr. Magellan[15] of London in 2 Vol's. octavo; it contains all the original work, with a great variety of useful additional information in the notes and in several additional sections.

The next work on Mineralogy is that of Wallerius[16] the Professor of Chymistry at Upsal. He published it a great many years ago in the French language but since that has republished it in Latin in 2 Vols., 8vo. All that I have to say with regard to this work is that if a student is to be confined to one work it is the best, on account of the excellent descriptions of the Species and Varieties of Fossils it contains.

The next work is a very small one but one of prodigious merit, the Sciagraphia of Bergman.[17] He was a man advanced in Life before he applied himself to Chymistry; his studies in natural history previous to that were principally confined to Botany and Entomology. The works of Bergman plainly shews the superior nature of his genius, for in point of precision he was the first of all Chymists. Accordingly it appeared greatly in this little work; he was particularly engaged in the discovery of the primitive earths and he fix'd them to be five. Ist The Calcareous earth; IId Argillaceous, formerly termed the earth of Alum; IIId the earth of Magnesia; The IVth is the *Terra ponderosa;* and the Vth Siliceous earth.

He discovered these and after all his trials he informs us that he never discovered any other whatever, and farther that he never was able to convert any one of them into another.

The last work I have to mention is the Elements of Mineralogy by Mr. Kirwan.[18] This work proceeds entirely on the plan of Bergman, greatly enlarged however by experiments of his own and all the discoveries made by other writers since Bergman published.

Before I go farther it may be here proper to give a general account of the primitive earths of the fossil kingdom.

It is very clear that the primitive matter of fossils forms the most proper Characters of the Classes, and that they ought to be arranged according to their composition. In consequence of this therefore the primitive earths lay a foundation for the most natural Characters. But to detect these is by no

---

[14] *An Essay toward a System of Mineralogy,* translated by G. Engerstroom with additional notes by E. Mendes da Costa (London, 1770).

[15] J. H. de Magellan (1788).

[16] J. G. Wallerius, *Mineralogie.*

[17] Tobern Bergman, *Sciagraphie. . . .*

[18] R. Kirwan, *Elements of Mineralogy.*

means so easy, and notwithstanding the labours of Bergman they are as yet involved in obscurity and it is highly probable that a future period may put them in a very different light than what they are at present.

The famous Stahl supposed them to be 2 in number, the Alkaline and Vitresent earths.

Ferber[19] thought they all consisted of Argillaceous and Calcareous earths.

Mr. Baumè[20] likewise entered into Stahl's idea and thought there were only the Alkaline and Vitrescent, and he considered the Alkaline as the only primogenial earth.

There are others that have fix'd the primitive earths as three; the famous Becher was the first who used that sort of triple division and these were the Calcareous, the Vitrescent and Apyrous earths. By the last of these he comprehends that earth of which Amianthus, Asbestus, Talc etc. are composed.

Linnæus followed this division, and likewise Wallerius and Justi.[21] Linnæus ascribed it to his countryman Bromel[22] but it unquestionably belongs to Becher. Lately however Wallerius seems inclined to think that the primitive earths are only two, the Calcareous and Argillaceous.

We have likewise a triple division given by Scopoli, the Calcareous, Argillaceous and Siliceous, and he appears to be of opinion that the Siliceous is only primogenial earth.

Pott and Wolsterdorf fix't them at 4, the Argillaceous the Alkaline the Gypseous and Vitrescent.

Of late Mr. Fourcroy has fix't them at 4. The 1st of these he calls the Gemmeous earth, which he distinguishes from that of Quartz or of Chrystals and the distinction is certainly well founded. It is chiefly this, that all the Quartz Crystalline earths are fusible by the fossil alkali, which is not the case with the Gemmeous earths for these are only fusible by Borax or the Microcosmic salt. In whatever order fossils shall afterwards be arranged we are certainly much obliged to Mr. Fourcroy for this distinction.

Newman fix'd the primitive earths to be 5, the Argillaceous, the Calcareous the Gypseous, Crystalline and what he calls the Talcy earth, by it he means that earth of which Talc, Asbestus, Mica and Amianthus are composed.

Cronstedt fix'd them at no less than 9.

    Ist.   The Calcereous earth, comprehending all Chalk, Lime, Marble, etc.
    IIdly.   The Siliceous earth, of which all Chrystals are compos'd.

---

[19] Johann H. Ferber, translator of Bergman.

[20] Antoine Baumé (1728–1804).   [21] J. H. G. Justi, *Mineralreichs.*

[22] Magn. von Bromel, *Inledning til Nödig Kundskap om Bergarter, Mineralier, Metaller, samt Fossilier* (Stockholm, 1730).

III.   Garnate earth, of which the Garnates are compos'd.
IV.   Argillaceous, to this belong the Clays and all stones formed from them.
V.   Micaceous earths, containing all the Micas.
VI.   Flux Spat earths.
VII.   Amianthine earths.
VIII.   Zeolitical earth.
IX.   Earth of Manganese.

But as I have already observed Bergman fix'd them to be five, the heavy earth, the Calcareous, the earth of Magnesia, the earth of Alum and the Siliceous earth.

This fivefold division is unquestionably the most just which has been yet attempted, and it is remarkable that it brings us back to the place where Stahl left us. He considered the primitive earths as of two kinds and Bergman's 5 earths amount to no more.

Technical terms in mineralogy defined, (vide Syllabus) and exemplified by Specimens.

In Mineralogy there are two different methods of Arrangement. The one founded on their Natural or external Characters; the other on their internal or Chymical properties, the one is the Natural the other the Chymical method.

After many trials it appears now pretty evident that there cannot be an arrangement of fossils founded merely on their external characters. At the same time it likewise appears that there can neither be an arrangement formed merely Chymical, for tho this is undoubtedly of great use in the science of Chymistry yet it is exceedingly unfit for the purposes of a naturalist, and it is even very questionable if all the Genera, Species and Varieties of fossils could be distinguish'd by Chymical Characters.

But the characters of the Classes and Orders must be formed on Chymical principles, while those of the Genera, Species and Varieties seem to depend chiefly on their natural and external marks.

The methods observed in the Syllabus in your hands is properly neither one nor other of these but a mix'd method, compounded of both.

It contains a greater number of Classes, Orders, and Genera than there are in any other System yet published and it is now acknowledged that the number of fossils is so large that the different members of the System must be enlarged for their complete elucidation. The Classes are in number 19. That general arrangement of fossils so long and meritoriously used is here adopted. The whole fossil kingdom is here divided into Earths, Stones and

Minerals, under the last of these are comprehended the Salts, the Sulphurs and the Metallic bodies. Under these 5 divisions all the substances belonging to the fossil kingdom are clearly comprehended.

The first division, the Earths, is here distinguish'd as it must be in every natural arrangement, from the Stones, but this division I may here observe can have no place in a Chymical system; nor should it have any, because tho' they differ in external properties, they are at bottom and on their Chymical qualities the same.

As to the names of the Classes, Orders and especially those of the Genera, there are none of them new and none of them mine. I have been at pains to select them from the writings of the older authors and have chosen such as appeared to me most fit for the purpose. The number of Genera is exceedingly extensive and it is proper that each should have its name, for this purpose many of the Classical names from Pliny are adopted, altho' in every case he did not use it to express precisely the same substance, but some one nearly allied to it.

The System you have in your hands contains 322 Genera. Under these are comprehended all the fossils that I have ever seen in nature, that I have ever seen in collections, or described in the writings of Naturalists or Chymists. Since the year 1782 I have been able to add but 3 or 4 Genera tho' the new genera will now and then occur.

CLASSIS FOSSILIUM

The great depth of the former arrangement of fossils arises from the confined numbers of the divisions, as they consist of too few classes, orders, and genera; and I am convinced, that, until the members of the mineral system be augmented, we will never obtain a thorough knowledge of the species. On this principle I have given a larger number of these than has been hitherto given by mineralogists.[23]

There are two general methods in arrangements: The first strictly chymical; the second depends on the natural characters of the bodies drawn from their external appearance; and to these may be joined a third which is mixed; partly composed of both; which last I reckon the best for our own use, and which I will follow, receiving no chymical characters but those which are absolutely necessary. The others will be drawn from the natural appearance of the fossil.

[23] The introductory paragraphs are from D.C. 2-19.

Then I have put into your hands the

SCHEDIASMA FOSSILIUM

which I have compiled for your use. But you will observe all fossils are divided into eighteen classes, the general arrangement of which is divided into three parts, vizt.

Earths
Stones
Minerals

This has been an old and established mode of division. All fossils on the globe belong to one or other of these three great divisions. The minerals are again divided into Salts, Sulphur, Mercury.[24]

REGNUM FOSSILE.

Clavis Classium.

Terræ,		1. Terræ.
		2. Calcaria.
		3. Gypsea.
		4. Phosphorea.
		5. Fusoria.
		6. Silicea.
Lapides,		7. Steatitica.
		8. Apyra.
		9. Zeolitica.
		10. Micacea.
		11. Petræ.
		12. Saxa.
		13. Concreta.
MINERALIA.	Sal.	14. Salia.
	Sulphur,	15. Inflammabilia.
		16. Pyritæ.
	Mercurius.	17. Semimetalla.
		18. Metalla.

[24] An incomplete classification follows in D.C. 2-19. In its place I have included the earlier (1781) *Schediasma Fossilium*. Interested readers may wish to compare this with the outline of Walker's 1787 *Classis Fossilium* printed by Jameson in his *System of Mineralogy* (1820), pp. xxiii–xxiv.

CLASSIS I

Terræ

ORDO I.
FIGULINÆ.

  Argilla
  Bolus
  Cimolia
  Lutum

ORDO II.
FIMOSÆ.

  Humus
  Limus
  Turfa

ORDO III.
CALCARIÆ.

  Creta
  Marga
  Torvena
  Parætonium
  Acudema

ORDO IV.
GYPSEÆ.

  Morochtus
  Puteolana

ORDO V.
SILICEÆ.

  Segullum
  Tasconium

ORDO VI.
ASPERÆ.

  Tripela
  Velitis

ORDO VII.
LAPIDEÆ.

  Arena
  Glaræa
  Saburra
  Sabulum

ORDO VIII.
SALITÆ.

  Magnesia
  Calamita

ORDO IX.
INFLAMMABILES.

  Egula
  Alana
  Umbrica
  Mottena
  Liparæus

ORDO X.
METALLICÆ.

  Ochra
  Miltos
  Melanteria
  Amphitane
  Chrysocolla
  Cerussa
  Helcysma
  Melia
  Sinopis
  Zaphara
  Sandyches

CLASSIS II

Calcaria

ORDO I.
CÆMENTARIA.

  Psadurium
  Mennonia
  Orobias
  Meconites
  Conissala

ORDO II.
DÆDALEA.

  Marmor
  Chernites
  Graphida

ORDO III.
STIRIACEA.

  Stalactites
  Phengites
  Stalagma
  Poros
  Undulago
  Osteocolla
  Phacites

ORDO IV.
SPATA.

Amorpha
  Castine
  Saurites
  Trichestrum

Figurata
  Rhombites
  Paropsis
  Halotessera

Hyodon
Drusa
Sprenus
Eristalis

CLASSIS III.

GYPSEA.

ORDO I.
PLASTICA.
  Alabastrum
  Emites
  Elasmis
  Inolithus

Lachnis
Hepatites

ORDO II.
SELENITICA.

  Halosanthus

Specularis
Calgum
Cachimia
Acenteta
Astrapia
Sagda

CLASSIS IV.

PHOSPHOREA.

Amorpha
  Lithosphorus
  Asyctos

Figurata
  Chrysolampis
  Zizaca

CLASSIS V.

FUSORIA.

ORDO I.
AMANDINA.
  Amorpha
  Myrmecias
  Plasma

Figurata
  Carystius
  Omphax

ORDO II.
GARAMANTICA.
  Sandastrum
  Sapinus
  Sacodion
  Soranus
  Turmalina
  Chrostasima

CLASSIS VI.

Silicea.

ORDO I.
QUARTZOSA.

Petridium
Galaciocos
Fibraria
Androdamas

ORDO II.
JASPIDEA.

Ærizuza
Xanthus
Tanus
Nephriticus
Vitraria
Sinpolia
Lazuli
Azurium
Malachites?
Turchesia?

ORDO III.
LITHIDIA.

Substantia uniformis:
Unicoloria.

Pyromachus
Homochroa
Scrupus
Hirundinaria
Chalcedonius
Sarda

Substantia difformis:
Maculata.

Achates
Phytomorphos
Stigmates
Galaxia

Polyzonia.

Calculus
Capnias
Onyx
Sarcites
Camea

Substantia uniformis:
Versicoloria.

Opalus
Pæderos
Asteria
Scambia
Hydrophanes
Cacholonius
Pramnion

ORDO IV.
GEMMÆ.

Hyalodes
Etindros
Lyncurion
Beryllus
Amethystus
Chrysolithus
Smaragdus
Topazius
Sapphirus
Rubinus
Adamas

CLASSIS VII.

Steatitica.

ORDO I.
SAPONACEA.

Smectis
Leucogæa
Colubrinus

ORDO II.
OLLARIA.

Siphnius
Catochites
Callaica

CLASSIS VIII.

Apyra.

ORDO I.
AMIANTINA.

Filamentosa
Spartopolia
Aluta

ORDO II.
ASBESTINA.

Fibrosa
Tricheria
Savinus

CLASSIS IX.

Zeolitica.

Amorpha
Nisuros
Anacites

Figurata
Bostrychites
Chalazias

CLASSIS X.

Micacea.

Argyrites
Chrysophis

Aspilates
Bractearia

Talcum
Ammochrysos

CLASSIS XI.

Petræ.

ORDO I.
OPHITICÆ.

Echidna
Cyamea
Nicomia
Lydius

ORDO II.
QUADRINÆ.

Ovadrum

Trapezum
Phloginus
Tephria
Cysteolithus

ORDO III.
COTACEÆ.

Passernice
Æcopis

Cherile
Thyites
Pharmacitis

ORDO IV.
SCHISTOSÆ.

Steganium
Ardesia
Tænarium

CLASSIS XII.

Saxa.

ORDO I.
CALCARIA.

Larbason
Aconis

ORDO II.
ARENARIA.

Sympexium
Efestis
Filtrum
Molaris
Sarnius
Scyrus
Blotta

ORDO III.
PORPHYRIA.

Leucostictos
Lithozugium
Baroptenus
Gasidanes

ORDO IV.
SCHISTOSA.

Basaltes
Sideropoecilon
Pardalion

ORDO V.
AMANDINA.

Calomachus

ORDO VI.
GRANITICA.

Syenites
Psaronium

ORDO VII.
MICACEA.

Lepidotes
Naxius

ORDO VIII.
ASBESTINA.

Amython

ORDO IX.
METALLICA.

Stomoma
Paneros

CLASSIS XIII.

Concreta.

ORDO I.
TERRESTRIA.

Geodes
Cissites
Ætites
Enhydros
Septaria
Stelechites
Taphiusa

ORDO II.
AQUEA.

Eumeces
Tophus
Glaphyrum
Peridonius

ORDO III.
IGNEA.

Lava
Scoria
Pumex
Cinis
Pompholyx

ORDO IV.
METALLICA.

Graptolithus

CLASSIS XIV.

SALIA.

ORDO I.
ACIDA.

Vitriolicum
Muriaticum
Nitrosum
Spatosum

ORDO II.
ALCALINA.

Natron
Halmiraga

ORDO III.
ACIDO-ALCALINA.

Arrhenicum
Muria
Nitrum
Borax
Aprium
Ammoniacum

ORDO IV.
ACIDA-TERREA.

Alumen
Aphroselinum

Psoricum
Polytrix

ORDO V.
ALCALINA-TERREA.

Calastræum
Halosachne

ORDO VI.
ACIDO-METALLICA

Sideranthos
Chalcanthum
Cadmia

CLASSIS XV.

INFLAMMABILIA.

ORDO I.
AERIA.

Mephitis

ORDO II.
PHLOGISTICA.

Sulphur

ORDO III.
BITUMINA.

Naphtha
Petroleum
Maltha
Mumia
Asphaltum

ORDO IV.
CARBONARIA

Anthracion
Bena
Ampelitis

ORDO V.
ELECTRICA

Gagates
Succinum
Ambra

CLASSIS XVI

Pyritæ

ORDO I.
SULPHUREÆ

Pyrobolus
Hypestionus
Marcasita
Placodes

ORDO II.
ARSENICALES

Leucopoecilos
Auripigmentum

ORDO III.
FERREÆ

Smiris
Syderea

Megalea
Magnes

ORDO IV.
AMANDINÆ

Molybdæna

CLASSIS XVII

Semimetalla

ORDO I.
ARSENICALIA

Arsenicum
Niccolum
Cobaltum
Vismutum

ORDO II.
SULPHUREA

Stibium
Zincum

ORDO III.
FLUIDA

Hydrargyrum

ORDO IV.
DUBIA

Platina

CLASSIS XVIII

Metalla

ORDO I.
DURA

Ferrum
Cuprum

ORDO II.
FLEXILIA

Plumbum
Stannum

ORDO III.
FIXA

Argentum
Aurum

DELINEATIO FOSSILIUM

The Pamphlet[25] now put into your hands is the first attempt to give you a description of the terms used in mineralogy.

Though mineralogy has been cultivated in all ages, still it lies under the disadvantage that the terms of the science are unfixed. There is the greatest necessity that this ought first to be done in any course to be given in mineralogy. This has

[25] The glossary that follows comprises lecture 25 in D.C. 2-18.

been to me a work of great labour. I have perused all the authors on it, and have pick't out these terms from them, choosing the most eligible, and rejecting many that I termed exceptionable, and used the synonima of authors.

Many of the terms appear new and uncouth to a classical mineralogist. But I have added little or next to nothing of my own. I have reduced the whole Terma under the seven general heads mentioned in the Pamphlet. They compleatly comprehend all the terms that have been used or that are proper to be made use of in mineralogy. They are

1. Situm .... Situation
2. Substantium .... Substance or Composition
3. Figuram .... Figure or Appearance
4. Structuram .... Structure
5. Partes .... Parts or Members
6. Consistentiam .... Consistence
7. Qualitates .... Qualities

As to the first:

Situs

Comprehends those terms that respect the situation of fossils.[26]

Stratum—A bed or layer.

Stratulum—A diminutive or small stratum remarkable short or thin.

Interstratum—That space sometimes found between two strata, as is applicable to many fossils.

Stratosum—Is applicable to every fossil found in a stratum.

Rectum—Is where the sides are in a straight line.

Undulatum—One or both sides waved.

Depressum—One sunk in the earth and not appearing.

Emergens—Where it appears at what miners call days or breaks forth at the surface of the Earth.

Inclinatio—The inclination or dip; the angle or degree at which strata are inclined to the horizon.

Horizontale—Horizontal stratum; one which is perfectly or nearly parallel to the horizon.

Obliquum—Is one inclined to the horizon at a small degree from 0° to an angle of 30°.

Elevatum—Still more inclined to the horizon than the former, at an angle from 60° to near 90°. This last is seldom to be met with. I never saw strata right angular, and I never found a stratum above 85°.

Directio—This means the direction or stretch of the stratum as it runs lengthways and is called the streik [strike] in this country.

[26] As elsewhere in these lectures, anything dug from the ground. Here, specifically, minerals.

Mineralogy

TEGMEN TERRÆ—Is that superficial stratum which serves every where as a covering to the earth. The staples of the earth, or the vegetable earth crust or soil.

RUPES—As distinguished from stratum. By it we understand the mountain rocks, distinguished by their inclination from horizontal covers or stratified matter. This is conspicuous in them, that they are always almost vertical from 60 to 85 degrees.

ERECTA—Where they appear upright and are very frequent in mountainous countries.

COACERVATA—As where there are great bodies of rocks of considerable height without any visible interstices, and is sometimes seen in the same place.

RUPESTRE—Divided by Cronstedt into:

Petra—Simple Rocks.

Saxum—Compound Rocks.

When we use the word simple, it is applied to any rock in a limited sense and only in a comparative manner as there is no rock that can be considered a simple element. But simple rocks are those in which we do not observe any composition.

SAXA—Again as vice versa for they are rocks where the eye can observe a composition. This is conspicuous in marbles which are themselves divided into simple and compound. The white marble is the simple kind, the coloured and embroidered marbles are of the compound kind. I here produce to you

Specimens of the Compound Marbles

All compound marbles or stones have a *ground* and a *charge*. The terms:

PETROSUM—Inclining to simplicity more than to a compound rock or stone.

SAXOSUM—More compound than simple.

NODUS, NODULUS—A node or nodule is a rock found in a single solitary mass smaller than its surrounding parts.

Nodosum—Of or belonging to nodules.

Nodulosum—Of or belonging to small nodules.

PESSULUS—Signifies a rock or mineral interposed between two large strata or in the body of a vein. It is commonly harder than the strata. Here is a specimen which I produce to you. The English miners call it a *Barr* or *Bat* and the Scots miners a *Rib*.

Pessulatum—Of or belonging to Pessulus.

ANTRUM—A cave, cavern.

CRYPTA—A *Cavern.*

Cavernula—A smaller cavern.

VENA—A vein to be considered as the intermediate space betwixt two sides of vertical shale.

Venula—It is a diminution of Vena.

VENIGENUM—Is any fossil formed in a vein. This is applicable to many fossils which we would be otherwise at a loss to distinguish.

FISSURA—Are fissures or cracks in the earth which sometimes filled with matter different from the vein or surrounding strata.

SOLUTUM—Fossils found loose and detached.

PARASITIAM—Fossils formed on others, as all Stalactites.

SPORADICUM—Any fossil or stone included in any other stone, stratum, or rock; as common flint which is found in strata of chalk or limestone; agate, calcedony and cornelian rank here.

STALACTICUM—Any stone or fossil formed by the dropping of water.

SOLITARIUM—A single stone or fossil solitary by itself.

CONGESTUM—Any stones or fossils formed in a heap.

VAGUM—Any stone or fossil found detached from the original matrix.

SUPERFICIAL—Those fossils or stones found on the surface of others, as the a[r]borescent appearance on slates.

SUBAQUOSUM—Those fossils or stones found under water.

SUBCOSPETITUM—Those fossils or stones found under moss or turfs.

DIMENSIO—The length, breadth, and depth of different strata or rock, hence the terms . . . *stratorium, rupium.*

It is also necessary to describe fossils and stones according to their magnitude or size. Hence the best way for this is by comparing them with most common bodies as

MAGNITUDO—

> *Seminis papav[eris]*—The size of the poppy seed.
> *Pisi*—Size of pea.
> *Nucis avellanæ*—the size of nut.
> *Castaneæ*—of a chestnut.
> *Pomi*—of an apple.
> *Capitis humani*—Size of the head of a man.

MATRIX—Argillacea, calcaria, schistosa, quartosa, micea and concursio fossilium, eadem regione, *vel eodem loco* describenda.

Here we shall mark what fossils appear in the same place or any part of the earth.

EXTRANEA NOTANDA—as

> *Locus*—We in our own opinion ascertain where a mineral is found in what part of the earth. This is of great importance in both the vegetable and animal kingdoms but of still greater in the fossil kingdom.

§ 2 SUBSTANTIA

Are those terms of Mineralogy that relate to the mere substances. All the known minerals belong to one of the nine divisions that stand at the head of this arrangement. Vizt.

Mineralogy

1. TERRA
2. LAPIS
3. SAL
4. SULPHUR
5. BITUMEN
6. PYRITES
7. MINERA
8. SEMIMETALLUM
9. METALLUM

1. *Terrigenus*—This is a stone composed of earthen matter in a state of induration. Consequently it is perfectly opaque, and is never accompanied with any solvent or menstruum of any kind. And it is called geophams [?] on account of its pure earthy origin. It is mere earth in an indurated state.

2. *Fluctivagus*—This is a stone which has once been in a fluid state, or solvent in water or some menstruum. It is accompanied with some degree of transparency. As for example petrus silex, quartz and white marble.

Lapis

1. *Fluor*—Was anciently applied to stones of this class. Cronstedt has assumed it and taken its ancient sense.

2. *Aquatus*—Is a stone that has been in a state, not of solution but of diffusion, that has subsided and had once its discession [?] concreted as the tophir.

COMPOSITIO

Simplex—Simplicissimum.
Homogeneum—Is a stone composed of one kind of matter.
Heterogeneum—Is a stone composed of different kinds of matter.
Compositum—Is a stone composed of two or more simple substances.
Decompositum—Is a stone of two or more compounds or different kinds of matter.
Aggregatum
Sterile—If a stone contains nothing metallic, but if it does then it is called
Metalliferum
Chrystalisatum
Resolutio—Primitivum [———].
Detritum—Stones or fossils formed by the wearing away of others.
Cariosum—Fossils or stones having various substances or appearances.
Calciforme—Stones of the form or appearance of calx.
Precipitatum—A stone precipitated from a solution.

VESTITUS

Naturam—Native metals as virgin gold.
Purum—Pure.

Impurum—Impure.

Larvatum—Mixed and so disguised as not easily to be distinguished.

Terrificatum—This mixture of earth.

Lapidificatum—This mixed with stone.

Mineralisatum—This, when combined with sulphur or arsenic, which then put on the appearance of ores.

FOSSILS—are likewise called terreum, lapideum, salinum, sulphureum, bituminosum, pipiticosum, siliceum, micaceum, etc.

Materia prædominans, si fieri potest, præcipue statuenda.

It is of great consequence in Mineralogy to mark what principally predominates in a fossil.

§ 3 FIGURA

Comprehends those terms in Mineralogy that relate to the figure of fossils in general. These figures are of two kinds, vizt., *Indeterminate* and *Determinate*.

INDETERMINATA—AMORPHUM—A fossil of an indeterminate figure is so quite irregular, that it has no specific shape.

Fossile Amorphum—A shapeless fossil.

DETERMINATA—FIGURATUM—POLYMORPHUM—is where fossils have a fixt and regular figure. And hence the two terms figuratum, polymorphum. The first being fossils of a regular figure; by the second we understand a fossil of several regular figures, but of different kinds on the same fossil and sometimes of one shape and sometimes of another.

Figura—Adventitia vel Spontanea

Adventitia proceeds from adventitious external and mechanical causes. *Spontanea* is spontaneous and proceeds from some external principle or interior part of arrangement which is the case with all chrystalized bodies. So adventitious are called:

a. Plana, concava, convexa, teres

b. Lentiformis, reniformis, verrucosa, tuberosa, glandulosa

c. Spurica, hemispherica, ovata, subrotunda.

d. Scorieformis, dentata, bodroidea[?], dendritica, arborescens.

e. Stiriacea by dripping or running water—Stalactitical.

SPONTANEA—Fossils of a spontaneous figure. In these the figuration accidentally proceeds from an internal principle. They are:

Capillaria—or slender crystals resembling hairs.

Plumosa—Pennated like feathers.

Stellata—Star shaped.

Radiata—Radiated like the spokes of a wheel.

Cubica—Where of cubical shape.

Tessellata—With more than six regular sides, but on the whole roundish.

Rhombea—Rhombus shaped.

Rhomboidea—Approaching to that shape.

0

Cuneiformis—Wedge shaped, as some crystaline iron ores, but sometimes this wedge figure belongs to the adventitious form. Here is a specimen of that in the mountain schistus.

Prismatica—Prism shaped.

Pyramidata—Pyramid shaped.

CRYSTALLUS—This term has been always used as a term of art in Mineralogy, and improperly as a generic term, but is irregular and unscientific. It is therefore improper to use it as a genus, but to use it as a name.

Crystallinum—Columnaris—is what forms a column of some length.

ACAULIS—without a stalk and is uniformly composed of a pyramid.

AMPHIPYRAMIDATA—means a crystal composed of two pyramids joined with no column betwixt them.

BICUSPIDATA—Is a fossil which has a column and a pointed pyramid sometimes at both ends.

TRIGONA—TETRAGONA—POLYGONA—Respects the number of angles in a fossil.

TRIHEDRA, TETRAHEDRA AND POLYHEDRA—Respects the number of sides that are in a fossil.

COLUMNA—Has base sides and angles.

PYRAMIS—Has point sides and angles.

BASIS LATERA ANGULI

APEX LATERA ANGULI

RADIX—Is that part of a fossil which is fix't to the matrix.

§ 4 STRUCTURA

INDETERMINATA—Is where the structure has no sort of regularity; as in all the simple rocks.

SOLIDA—where the structure is round and solid.

FRUSTULOSA—Is where the fossils are shattery, full of fissures and easily break; but those fissures not very palpable.

FISSILIS—Is a stone that easily divides into longitudinal plates of a solid form.

POROSA—Is a fossil where the structure is porous.

FORAMINOSA—A fossil whose pores are of a larger size than the former in its natural structure.

From these are derived frustulum, fissura, porus, foramen, porulus.

SPONGIOSA—Relates to fossils that are not only porous, but very light and resembling a Sponge.

ARENATA—A fossil of a sandy structure.

GRANULATA—Is only a variety of the last and only a little larger than common sand in the grains.

ACERVATA—From these terms flow these words.

ARENULATA—*Granum, granulum.*

DECUSSATA—A fossil of a longish figure spread through the body or surface; that is, where the particle is longitudinal and irregularly spread thro the substance.

DETERMINATA—Is a regular figure in fossils.

SQUAMOSA—A fossil of a scaley structure resembling scales of fishes.

LAMELLOSA—A fossil divisible into lamellæ, as in all slates.

MEMBRANACEA—Divisible in like manner but flexible and sometimes elastic, as the talk and muscovy glass.

Squama Squamula
Lamella
Membrana Membranula } All are derivations from the last three described terms.

STRIATA—These are fossils whose parts are arranged in a lineal order in a parallel manner.

SPECIMEN—*Gypsum Striatum*

FIBROSA—Those fossils which are fibrous. The difference between [striata] and the fibrosa is this, that in the fibrosa the striæ are more easily separated than in the striata, and in this they can be divided like the fibres of a muscle. The asbestus is of this kind.

FILAMENTOSA—Resembling threads. These again afford the terms *stria, fibra, filamentum, fibrilla.*

VENOSA—Fossils that have veins, as the agate and onyx.

ZONARIA—These again resembling zones or belts, the veins larger and broader than in venosa, like ribbons.

STRATULOSA—Stratulated, or composed of a number of small strata.

CONCENTRICA—Such fossils as are composed of concentrated laminæ in their structure; as many sorts of agates, calcedonæ and onyx.

To these again relate the following terms: *vena, zona, stratulum, venula, zonula.*

RHOMBEA—Approaching to a rhomb in shape.

TESSELLATA

RADIATA

Hence these words, *rhombus, rhomboides, tessara, tesserula, radius.*

FRACTURA—By which is to be considered the surface of a stone when broken or its appearance at the fracture. This is of great use in mineralogy as it is sometimes plain, as in a slate.

Angulated—As in quartz.

Convex—Concave

Linnæus was the first who hit off this character. He observed all flinty fossils to break with a convex and concave surface, as may be observed in all flinty nodules.

I think this may be extended farther, and, in my opinion, may be owing to a minute laminated structure. Every gun-flint shows a concave and convex surface.

FIGURATIO—*Plana*—Plain.

 Angulata—Angulated.

 Convexa—Convex.

 Concava—Concave.

 Rimosa—Where fractures appear as if cracked.

 Variolosa—As if the fractures were pustulated.

 Pruinosa—Frosted, resembling hoarfrost.

 Undulata—Flexa.

 Polymorpha—Where a stone breaks with many facets or surfaces.

 From whence *unda, flexivia, faciecula.*

VARIEGATIO—Respects the colour and disposition of a fossil.

 Thus,

 Unicolor—Of one uniform ⎱
 Diversicolor—Of different ⎰ Colour

 Variegata—Variegated

 Terreformis—Earthy, powdery

 Hebes—Dull, not shining

 Vitrea—Glassy

 Catoptrica [sic]—Reflecting images, as if it were a mirror

 Punctata—Filled with points of particular colour

 Maculata—Stained

 Nebulosa—Clouded

 Fa[s]ciata—With regular lines

These four last relate to *Puntum, Macula, Nebula, Fascia, Punctulum, Nubeicula.*

 Micans—Too frequently confounded with micaceous, which regards the substance of mica, whereas *micans* represents the structure—meaning glistening.

 Pecten—Signifies the grain, grit or greet of a stone. It is either

 Uniformis or ⎱
 Difformis ⎰

 Æqualis or ⎱
 Inæqualis ⎰

 Manifestum ⎱
 Obscurum ⎬
 Impalpabile ⎰ Impalpable—where the grit is imperceptible to the naked eye; as in the crystalized gems.

 Pulverulentum

 Læve—Smooth

 Scabrum—Rough

 Asperum—Rougher

 Compactum

 Tenue—Fine

Crassum

Microscopii intuitus—Where it is necessary to examine the structure by the microscope.

§ 5TH PARTES

1. FRAGMENTA—The parts of a stone into which it is liable to be shivered by percussion or concussion.

> *Figura*—Thus, all quartzy fossils break with sharp angulated joints. Flinty fossils have all upon being broken either a convex or concave surface or both.

2. SUPERFICIES—Is either

> *Naturalis*—Natural or

> *Efflorescentia*—Covered with an efflorescence, acquired sometimes in the open air as all vitriolic substances.

> A *Structura*—Applicable to structure are:

> B *Qualitates*—Surface or superficies

3. TEGMEN—Is of three sorts

> *Tectum Nudum*
>
> *Terreum*
>
> *Lapideum*
>
> *Tenue*
>
> *Crassum*

> 1. *Velamen*—Is the thinnest fossil cover as in some flints on the sea shore.
> 2. *Cortex*—Is the thicker, and frequently seen in most sort of stones: as in the whin stones that have been long exposed to the air.
> 3. *Crusta*—This particularly belongs to flinty nodules which have not bark but crust which generally are concentric and is a number of crusts surrounding one another. From these we have the adjectives

> *Velatum*

> *Corticosum*

> *Crustosum*—And we should with attention mark this
>
> > a. *Substantia*
> > b. *Structura*
> > c. *Consistentia*
> > d. *Qualitates distinctionis inter fossilia eorumque tegmina sedulo perscrutari debent: Inde Lux e Tenebris.*

These coats consist for the most part, or indeed almost always, of the matter of a fossil in the state of decomposition, and this will lead us into just views of fossils when we meet with them in these circumstances.

4. GRANA—This means in point of size, as when small it is described by its size.

[5.] GRANULA—This differs only in size.

Particulæ Particellæ
    Crassa—Gross
    Subtilia—Fine
    Æqualia—Equal
    Inæqualia—Unequal
    Antica—Two or more kinds joined in equal portions or proportions
        throughout the whole.
  a. *Substantia*
  b. *Figura*
  c. *Structura*

6. Lamellæ Imbricatæ—Two coates covering one another like tyles on a roof.
    Concentriæ—Concentricated
    Dehiscentes—Parting
    Intercursantes—Crossing one another
  a. *Figura*
  b. *Structura*
  c. *Consistentia*
  d. *Qualitates*

7. Filamenta—Flexible like a thread and are either *paralella,* parallel, or *intertesta,*
  interwoven.
  a. *Figura*
  b. *Structura*
  c. *Consistentia*
  d. *Qualitates*

8. Fibræ—These are all rigid and ranged according to
    Dispositio—Their disposition
    Rectæ—*Paralellæ*
    Undulatæ—*Contortæ*—*Longitudinales*
    Transversæ—*Abruptæ*—*Discretæ*—disjoined
    Intertextæ—*Intersecantes*
    Fasciculatæ—Bundled—*decussatæ* spread
    Congestæ—In a heap—*sparsæ* asunder
    Figuræ—*Crassæ*—Coarse—*tenues*—fine—*capillares filiformes*
        Angulatæ—teretes—cylindrical
        Papposæ—*acerosæ*—chaffy
    Consistentia
        Rigida
        Fragilis
        Separabilis
        Inseparabilis
        Qualitates

9. Striæ ⎫
10. Tali ⎬ These terms are confined to *Eudus*[?] *helmontii,* a stone all divided
11. Septa ⎭ into cubical figures and these separated by partitions.

12. Nucleus—A stone whose central part is of different substance from its external parts or covering, as in the geodes. In the Eagle Stone the nucleus is loose and rattles and is then called
13. Callimus—This is an old term used by Agricola and signifies a metallic cover on any other fossil, and is applied to those fossils covered with a coat of pyritical matter, as some slates.
14. Armatura—Metallic covers on any other pyritical matter.

§ 6th Consistentia

Cohæsio
a. Solidum—A fossil possessing solidity.
Tenax—Tough
Liquidescens—which is a fossil liquifying in the fire
Fluidum—fluid; as some of the bituminous fossils
b. Cohærens—Adhering
Incohærens—Incoherent
Macrum—Dry to feel
Viscidum—Wettish. The three last are opposed to this—being earths of a clean dry substance.
c. Friabile—*Pulverulentum*
Inquinans—Staining as those stones stain the fingers upon touch.
Tersum—Those that do not stain the fingers
d. Ductile—*flexile—malleabile—fragile*
Elasticitas—*Elasticum—rigidum*
Saltans—A stone that bounds and springs, when let fall on a rock
Segne—A stone that does not bound on a fall
Scriptura—Tritura
Durities—
Molle—*durum. Inexpugnabile durissimum.* A stone that cannot be wrought underground without gun powder.
Scintillans—A stone that strikes fire with steel
Tenebrosum—A stone that does not strike fire on being struck with steel.
Scintillam silice—A flinty nodule, that gives fire on being struck against another of the same kind, but does not give fire to gun powder.
Politura—Respects the degree of polish of which a stone is capable.
   Arenaria—Sandy stone, marmorea, taspidea.
   Carnolica—*Crystallinæ, adamantina.*
   Poliendum—Polit[uram] habe[ns]. Is applied to a stone which is capable of some degree of polish.
   Nitida—A higher degree of polish.
   Catroplica—Such as we find in the best marbles, reflecting images.
   Polituram recusans—Incapable of any polish.

SCULPTURA—Respects the capability of being cut or carved. *Ungue,* with the nail— *Cultro*—the file.

>Lima rasile*—May be scraped with a chisel.

>Tornabile*—Capable of turning in a loom, as *Lapis olearis.*

>Cædum*—Cut with a chisel.

>Rebelle*—Refuses or resists chisels and falls off in irregular pieces.

>Chalybe sectile*—May be cut with steel.

>Sectura recta*—With a straight section, or *curvata,* a curved chip, like what arises from a board run over with a plain.

>Marmor*—As it cuts Marble.

>Vitrum*—Glass.

>Carnicolum*—Cornelian and scalpens.

>Limam respuens*—Rejecting chisel altogether.

CELATURA

>Sectile aqua—Arena—Tripela*

>Crystals—*

>Smirde*—Emery

>Adamante—*

PONDUS

>Natans*—Swiming, as shell marle, which is not reckoned a genuine fossil; as the *Subas mintanum*[?].

>Submergens Ponderosum*—A little heavier than water.

>Ponderis Metallici*—Of a metallic weight. There are few fossils which are more than three to one of weight with water, but contain some metallic mixture.

§ 7 QUALITATES

EXTERNE

COLOR

>Hyalinum*—Glassy colour

>Tinctum*—Some one or other of the prismatic colours

>Unicolor*

>Bicolor*

>Tricolor*

>Elementarium*—of fair colour

>Multicolor*—of many colours

>Diversicolor*—of different colours

>Varigatum*—Variegated

>Versicolor*—Changeable colours, as that of doves neck and in certain silks

>Rubrum*

>Aurantiacum*

>Luteum*

  Viride
  Cæruleum
  Purpureum
  Violaceum

Persistentia vel mutatio Coloris
  Aere
  Aqua
  Igne
  Stypticis

Pelluciditas
  Fulgidum—Very transparent
  Pellucidum—Less so
  Diaphanum—Still less transparent
  Obscurum—
  Surdidum—
  Cæcum—
  Opacum—
  Opacissimum—

Reflexio
  Hebes—
  Nitidum—
  Vitreum—
  Micans—Glistening
  Splendens—All shining

Refractio
  Reflexione vel refractione—Unicolor
  Diversicolor—
  Versicolor—
  Aere opacum—Opaque in Air
  Aqua pellucens—Shining in water
  Objecta duplicans—When object appears double as in the Iceland crystals.

Trituram—*Color pulvereus contusione*
  Attritu metallorum

Scriptura Concolor
  Alba—
  Cuna[?]—
  Atra etc.

Odor Naturalis Attritu* Igne. The natural odour by fire or friction as:
  Ambrosiacus—The ambergrease
  Fœtidus—*Suilus*—The shine stone
  Fragrans—Fragrant as naphtha

* *Scintillæ ex lapidibus sulphureis. Acutum ex arsenicalibus. Allia [acrem?] odorem spirant.*—Author's Note.

Mineralogy

Acutus—Sulphur—sulphureous

Suavis succinum—

Aliacens arsenicum—Arsenecal or garlic smell

Fistis asphaltum } The mephitic called by miners the bean blossom damp
Lethalis mephitis } from its resemblance to bean blossom smell

SAPOR

Linguæ austere adhærens. Those fossils which stick austerely to the tongue.

Ore Solvens—Dissolving in the mouth.

Linguam siccans—Remaining dry on the tongue.

Insipidus Gemma—Insipid, as in the precious stones.

Unctuosus Bolus—Unctuous, as in the bolus's.

Dulcis Arsenicum—Sweet as in arsenic.

Acutus muria—Sharp as in sea salt.

Frigidus Acris Nitrum—Cold and acrid as nitre.

Amaricans natron—Bitterish as in the fossil alkali.

Astringens alumen—Astringent as alum

Stypticus vitriolum—Caustic as the vitriol.

Nauseosus metallum—Nauseous as the metals, except gold and silver.

SUBSTANTIA

Tactus—In substances to the touch

Unctuosus—Unctuous

Lubricus—Lubricating, slippery

Macer—Dry and lean

Asper—Sharp

Superfic[i]es—In superfic[i]es

Scaber—Rough

Lævis—Smooth

Lenis—

Vitreus—Glassy

Olearius—Oily as the nephriticus lapis

Aridus—Dry as in the lapis pumex—pumice stone

AUDITUS—In its sound

Stridens—jarring

Clangosum—

Ærisonum—Sounding like brass when struck, or

Surdum }
Mutum } That gives little, or no sound at all. Nothing like clangor.

INTERNÆ

MAGNETISMUS

Intractable—not affected by the magnet.

Attractorium

Retractorium

Polos ostendens.

Pliny says that the Theamides repells iron, but this requires confirmation.

"Theamides omne ferrum abigit et respuit." Plin.

Electricitas attritu levia attrahens. Succinum drawing light bodies to them by friction—as amber.

Immersione aqua calida mastichen attrahens—Attritu adamas. On being immersed in hot water, drawing mastic towards it, as the diamond.

Oppositum genus electricitates ostendens. Showing positive electricity on one side, and negative on the other as the turmaline.

Turmalina

Phosphorus

Phosphorescens—"Nullus lapis phosphorescens nisi muria imbutus," Waller.

Attritu—Certain stones become phosphoric by attrition: as spatum, marmor, gypsum, adamas; as Pott tells us.

Ustulatione—By burning or heating as lithosphorus the bolognian stone.

Expositione radiis solaribus adamas crystallus: Other stones become phosphorescent by being [exposed] to the rays of the Sun, as the diamond and crystal.

Immersione aqua calida adamas. Others again by immersion in hot water, as the adamant.

Wallerius says that no stone will exhibit a phosphorescent appearance, unless it contains muriatic acid.

Origo

Generatio fossilis in loco quo repertum est, aut in alio, notanda. It is to be recommended to Mineralogists, that they be attentive in noticing the places where a fossil is to be found or in finding out from whence it came; and likewise to endeavour as far as possible to investigate the æra of its formation. Also, *æra formationis, si certis observationibus fieri potest delineanda.* [The age of formation must be delineated if possible by definite observations.]

a. *Genesis,* b. *Decompositio,* c. *Palingenesie regeneratione.* [Reproduce the life history.]

Usus—Ex observatione physica.

Character essentialis specificus, ex observatis, præcipue notandus. From what has been said, the essential specific characters may be found out and expected.

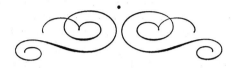

LETTERS

BETWEEN

John Walker and Charles Linnæus

LETTER I

To my Dear Sir, Linnæus, Knight of the Pole Star,
Physician in Service of the King of Sweden, and
Professor of Botany at Upsala, Sweden.

Since I venture, without permission, to write to you, I properly should ask your forgiveness; but you will, I hope, readily deign to forgive me, since my desire at the moment chiefly urges me to devote myself to that science to which you have so long given attention, and to which you have so successfully brought distinction.

For a long time I have taken an interest in the field of Botany and whatever hours I had found quite free from the interests that bear heavy responsibility, I have expended upon such inquiries, with great enjoyment, and, unless I am mistaken, not without advantage to myself. From your own writings I have received by far the most abundant enjoyment, and throughout several years scarcely a day has passed in which they have not given me pleasure. Consequently, I have eagerly sought to enjoy association with you by way of letters, with the hope that

This translation of the Latin letters, possibly by W. M. Smallwood, is in the Rare Manuscript Library, University of Edinburgh.

in this way there might be given the opportunity of acknowledging my debt to you; especially have I desired to share with you, as far as I may be able, whatever might serve your need. In order that I may not receive with an impassive brow your invaluable kindnesses, I not only acknowledge them frankly, but, if it is right to place your large services along with my small ones, I can only pledge equal service in return. Therefore, please assure yourself that I am not to allow even the least opportunity, wherein I may serve you, to pass by, and that I am to seize eagerly every shadow of kindness that may be shown to me. Whatever success can come from men whether in zeal or service, I shall deservedly and willingly give over to you.

The Society that was established in Edinburgh to promote philosophical studies, although it has faded for a long time, is, nevertheless, beginning, I hope, to bloom again; and in the future will employ greater diligence, in order that Natural Science may gain increase; and their own work will be brought more often into the light of day. I desire to know whether the volumes which this Society has made available to the public have come into your hands, or, if you have been without the opportunity of reading them, by what route they can be brought to Upsala. Furthermore, I ask you to inform me whether you wish ours to be the one of the Societies with which Linnæus deigns to associate himself.

The Society which I have just mentioned has decided to cultivate the study of Botany, but I fear that their purpose may fail of success, since this science lies so neglected among the Scots that, besides very many who are trained in a different kind of literature, scarcely one out of two hundred or more in the Edinburgh Academy who give attention to Medicine ever appears as a Botanist. This indifference has prevailed to such a degree that, as regards the bringing of distinction to this science, scarcely any one in Scotland has presented anything worthy of mention. Morison was, indeed, by birth a Scot, but only in this fact may we feel pride; he, however, in order that his unfruitful land might not lose the hope which it had gathered concerning his ability, quickly went away to foreign peoples, which received advantages from his works, and glory from the rewards that were conferred upon him. Houston, it is true, met his death at an hour inauspicious for Botany. These alone of our countrymen have accomplished something in this field of literature.

It is more to be regretted that in Scotland there is a lack of success in Botany, because of the very many professors of Medicine, who go from Scotland to all regions of the world; to these frequently are given suitable opportunities of cultivating the science, as it must be. But, since they had neglected to give attention to Botany before going abroad, the many excellent opportunities of this kind are valued lightly, and their loss is to be regretted.

Because of my regard for Botany I can scarcely endure with a calm spirit its present neglect. I have desired to put an end to this neglect, and I have determined to ascertain whether it is possible. Impelled by this purpose, I seem to feel that the best plan is to present this science briefly and to make it easier at the beginning, in order that by this plan the hearts of the young may burn with desire to learn Botany. And, furthermore, nothing could better contribute toward the attainment of this end than catalogues carefully compiled of foreign and native plants. This is apparent from the general agreement of Botanists, since by far the most numerous classes of Botanists are the "Adonistæ" and "Floristæ."[1]

Gardens abounding in foreign plants are most useful seminaries for the training of Botanists; to such gardens our country provides us with very limited access. This defect has risen from the fact that thus far in our country there has been a lack of Botanists endowed with experience and ability, to enlarge and meet the expense of such gardens. Accordingly, this defect must first be removed; Botanists must be obtained before we can find the gardens; and since at present we can not expect our plan to succeed from gardens, it is necessary that we attempt a different way; we must seek a different method, according to which the hearts of the young may be inspired by this study. From what source, I ask you, are the gardens, so essential and adapted to this science? From what source, unless students of this science can recognize to their complete satisfaction, without difficult and uncertain scrutiny, whatever plants they may wish? The same end is attained, and the same advantages are enjoyed by those who are learning from the accurately listed Flora of the region where they are sojourning; but we are equally destitute of Floras and Adonides.[2] Nothing of this kind is superior to the *Compendium* of Rajus[3] that has

---

[1] These are the terms used by Linnaeus in his *Bibliotheca botanica* (1736) to designate the classes of authors who compiled catalogues of the exotic plants in a botanic garden or of the native plants in a limited geographical region, respectively. *Bibliotheca botanica*, p. 64, reads (in translation): "*Adonistae* is my name for Botanic Collectors who have catalogued exotic plants that have been grown (from seed) in or transplanted into some Garden. Since exotic plants can be studied or described much less commodiously from dried specimens than in their natural habitat, enthusiasts of the art have established Botanic Gardens to enable visitors to fix exotic species in their memories." And again, p. 84: "*Floristae* are those who have provided catalogues of practically all the plants which grow in a definite and limited region. They have called their books by so many titles that their treatises hardly ever have the same title. Hence from these numerous synonyms I have chosen the one that is best, simplest, most appropriate, and most familiar, that is, *Flora*." Linnaeus' own catalogue of the plants native to Sweden is an excellent example of the type, and he entitled it *Flora svecica* (1745, 2d ed. 1755), as if the Roman goddess of flowers (Flora), transported to Sweden, were listing her possessions there. He also provided the outstanding example of the other class in his *Hortus Cliffortianus* (1737), an elaborate catalogue of all the plants growing in George Clifford's garden near Amsterdam or preserved as dried specimens in his herbarium. The prefatory matter of this volume includes a catalogue of Clifford's botanical library, arranged under the same headings as in the *Bibliotheca botanica*. (For a sample page and brief commentary, see *Huntia*, 1 [1964]: 57–58.) The titles listed under the two headings there, about twenty in each case, do not outnumber those from some other classes, but in the succeeding twenty-five years such books did in fact multiply among contemporary botanists, largely as the result of Linnaeus' precept and example. [Communication from J. L. Heller, Department of the Classics, University of Illinois.]

[2] That is, we are equally lacking in catalogues of our indigenous plants and in botanic gardens for exotics. The name *Adonistae* was coined by Linnaeus from the term *Adonides*

been brought into the light of day among us; this, although you rightly extolled it with relative praise, is, nevertheless, full of errors, and at the same time it suffers from defects and non-essentials. Since his method is generally unfamiliar, a double task confronts one: to find his plants and to investigate them in other methods. Nor is this Compendium at all suitable for Scotland, since the testimony of the authors relates only to regions of England.

In all, only a very small part of Scotland thus far has been surveyed with the eye of a Botanist, nor has any one by visiting this region carried his testimony to the point of being able to list its plants. Such a task is more to be desired than to be hoped for. But to begin this, and to render this study of Botany easier for the students in this Academy, it was my plan to publish a Flora of the native plants that grow around Edinburgh. To this task I was invited by many, but the one who first induced me to attempt it and particularly increased my interest, was William Cullen, M.D., Professor of Chemistry in the Edinburgh Academy, with high experience in his profession and in the practice of Medicine, and deservedly famous in our part of the world, and also a profound teacher of all Natural History. And since I have decided to follow the sex-system, I thought that it was best to consult the author of it. I wish there to appear in this Flora a drawing of each genus, prefixed to the species, but I am uncertain whether whole drawings, or only the principal and distinctive parts, in the same manner in which they are closely arranged in Nature's system. I have heard that you wish to bring again into the light your *Species of Plants,* and, if drawings of the *genera* are to be added, I have decided to postpone my own purpose until I may enjoy the benefits of this work of yours. To this *Flora* I have decided to add a catalogue of the Submarine Zoophytes, of which very many are found on the nearest shores. Accordingly, I have chosen this opportunity of leading those who are interested in the science of Botany to a knowledge of these sea-offsprings; not a few have the opportunity in warmer regions to examine the Zoophytes while these are still alive. In the history of these, much is lacking; since thus far they have been examined in carefully preserved specimens. Although animal life plays the principal part in these mixed bodies, nevertheless, when they are arranged in order, they should, I think, be considered plants rather than animals, because their species can best be investigated from their ramification and external appearance, rather than from the microscopic

---

which had been applied, with a similar allusion to classical mythology, to botanical gardens in general. See *Bibliotheca botanica,* p. 65 (translated): "*Adonis* is the name for a garden which lovingly rears exotic plants that have been sought from all parts of the earth and is equipped with a hothouse, or home of Adonis, to prevent the loss of the growing plants from frost." [J. L. H.]

[3] The third edition (by Dillenius) of John Ray's *Synopsis methodica stirpium britannicarum* (1724),which Linnaeus had cited in the *Bibliotheca botanica* (p. 84) as an outstanding Flora. See also the page reproduced in *Huntia* (cited above) from the *Hortus Cliffortianus,* where Ray's book is listed (no. 141) with the fitting comment, *Flora inter omnes prima censenda.* [J. L. H.]

living things that constitute them and dwell in them. It is not my plan to give a history of plants which I wish to classify, but merely to show very briefly their species. Therefore, I think that from these few there will be some on which different synonyms should be placed than those of the *Species of Plants* and of Rajus' *Compendium;* these species we assume are from the books of our student; each of these species have different synonyms, and especially those of Pinax (Pinacis), who is, as it were, a link between all authors. My plants will almost entirely consist of those which have long been discovered, and I can hope that only few new ones will be attested, except in th *Cryptomagia.*

Concerning these matters, I ask, as a very great kindness, your advice, or suggestions that can give as accurately and fully as possible the established Flora. Nor may I despair concerning my request, since I am convinced that you are deeply devoted not only to the science of Botany, but also to all who are engaged in these studies. Whatever may be your wish, it will be great as a law to me; whatever you may ask, I shall think that it is a command to me; whatever may be your command, I shall carry it out even more promptly than I could carry out my own wishes. I am writing this to you, but modestly and hesitantly, for I weighed in my mind whether I should send this letter to you, or put away my thoughts in silence; for I do not wish to take advantage of your courtesy, I do not wish to take from your public duties the time while you would give anxious thought to my personal interests; I do not wish to weary you, or to filch away the time that would be better spent.

In the field of Botany you have seized, in advance of all, the palm of victory, so that you have ascended a pinnacle to which the discouraged hearts of others never attain. Your own genius has opened this path of glory, it has carried ahead the torch, it has lifted the signal, it has laid firm foundations for your honors. In the wide theater of the whole world you are now worthily engaged, even while you are beset by so many duties—although you are always so busy, and spend all your days and hours on the business of the Republic of Letters. Against the advantage of this Republic of Letters I seem to commit a fault, when I venture to call you away by my letters, while you are intent upon more weighty matters. If, indeed, I find you indulgent to me, if you have sufficient leisure apart from your grave responsibilities so that you do not disdeign to read this letter and to send an answer, neither my hope nor my expectation will disappoint me, and I shall think that henceforth I am deeply obligated to you by this kindness.

You have the whole circumstance that impelled me to write. I went farther than I had planned at the beginning; nothing delights me more than to write to the one whose absent self I always carry in my eyes. There is nothing more for me to write, except to spur you on in your course, and to urge you to be willing to bless the learned world forthwith by your writings, which all are eagerly awaiting.

Farewell, illustrious Sir. May you prosper in your undertakings, and may you keep unimpaired your kindness to me.

I am writing at Woodhouselee,
near Edinburgh, January 8, 1762.
P.S.

[Drawings of plants]

Character of *Bryum striatum* (*Species of Plants,* p. 1115, n. 2)
Male flower

terminal, solitary, sessile.

Filaments	. . . . .	eight, compressed, connivent, of which the four in the middle are shorter.
Anthers	. . . . .	eight, blackish purple, sub-rotund, incumbent, sprinkled with purple powder.
Receptacle	. . . . .	foliaceous.

Female flower

terminal, solitary, on a different plant.

Peduncle	. . . . .	very short, filiform.
Calyx	. . . . .	calyptra conical, striate and pale.
Pistil	. . . . .	ovary ovate . . . Opercular style deciduous, spherical at base, filiform toward the apex . . . Stigma acute.
Pericarp	. . . . .	Capsule ovate, uniloculate, the mouth with slender teeth.
Seed	. . . . .	powder-like.
Receptacle	. . . . .	foliaceous.

If you favor me with a reply direct, if you please, by way of the post,

Mr. John Walker,
Minister of the Church of Glencorse,
at Edinburgh,
en Ecosse

LETTER II

Charles Linnæus, Knight, sends greetings to John Walker,
a most distinguished gentleman.

For a long time now I have been grieved because, after my good friend, I'sac Lawson died, I have not found in all Scotland a friend with whom I could discuss the plants of that region, and by whom I could be instructed in them. And so with happy omen you, honored Sir, have come to my aid. Therefore, for both reasons, I eagerly receive your great service to me, and I acknowledge it with gratitude.

There can be no doubt that Scotland will be found to rejoice in a great many very rare plants, not in the list of Rajus, nor in other accessible works, since your region abounds in very high Alpine. mountains, forests, and waters. For many years I have left no stone unturned, that I might obtain one perfect specimen of the *Erica Daboecii,* which is the *Erica Cantabrica,* with a very large flower, Tournef. *Erica Dabeii Hibernis;* Raj., *dendr.* 98; *Erica Hibernica,* Pat. *Gaz.* I. 2y, f4, (Petiver, Gazophylacium [sic].) Nor have I been able to obtain it; it is, to be sure, a plant of Ireland, and is not a native of Scotland; but I do not doubt that it is even found in your Alpine mountains. If you meet with it, please send to me a twig with a flower, in a letter; please make a good sketch and explain it properly. If ever there were a Flora most eagerly desired, surely such an one would turn out to be Scottish. I wish that you would finish this in a short time, and while I am still living. Scarcely should I welcome another more greedily. I beg you to add for each of your plants not only its place, but also its soil, in order that they may more easily be cultivated in gardens, while the seeds are being obtained.

I have read the *Edinburgh Proceedings,* which are for physicians the most excellent proceedings of all the learned Societies. In them I have found but few things pertaining to Botany, except the species of *Caldenii* and *Hyperici.*

You ask whether I should wish to be received into this famous Society; I do not see how it could be done as long as Cl———— is a member, who has black-balled me.

If some rarer and less known plants should come to your notice, if you should send them to me, I shall examine them and frankly write to you my opinion about them.

The second edition of the *Species of Plants* is now in preparation, as all the copies of the previous edition have been scattered,—increased by more than 1000 plants and a great many observations. When that edition is finished, I shall publish *The Queen's Museum,* with descriptions of Indian insects of the Indes and of other foreign animals.

With amazement I read your description of the *Bryum Striatum,* and with great impatience I am awaiting the first days of spring, when I can search for the male flowers of this *Bryum,* which you first discovered and thus anticipated me; surely if I shall find in it the eight named *stamina,* that moss with your observation, will unlock a new chamber of Nature, through which we shall enter a hitherto hidden palace of Nature; nor will there be a doubt that we shall have similar success in the other mosses. If this should happen, as I ought not to doubt your eyes, surely it would bring to you immortal fame. As a result of this one experiment I do not doubt that you are one of the keenest Botanists.

Continue, as you have begun, and enter the hidden places of Nature, and conquer new kingdoms there; and, furthermore, keep in your heart your love for me.

Upsala, February 22, 1762.

If you write in reply, address your letter to The Royal Society of Sciences of Upsala; I open all its letters.

I do not doubt that you can wonderfully enlarge the history of Molluscs, Zoophytes, Lythophytes, even now incomplete. I wish that you would undertake this.

To
    Mr. John Walker,
    Minister of the Church of Glencorse,
    Edinburgh, en Ecosse.

LETTER III

John Walker sends greeting to Mr. Charles Linnæus,
a famous gentleman and Golden Knight.

I have received your letter written at Upsala, a most welcome letter, since I learned from it that you are well. With very great pleasure I welcome this opportunity to inform you, at the request of the Edinburgh Society, that on June 17 you were unanimously elected to membership. On the same day John, Count de Bute, and his brother, Mr. Steuart McKenzie, natives and ornaments of Scotland: one, to-day governor of the whole British Empire; the other, envoy to the king of Sardinia; these were received into this Society.

I have explored the Scotch mountains this summer in quest of *Erica Hibernica,* but thus far to no purpose. Ireland at this time provides no one experienced in the field of Botany. I have written to the Bishop of Ossory,[4] and to a physician in the town of Dublin; and, if the matter turns out well, you will hear later.

In this packet of letters you will find specimens of paper, and of the plant from which it is made. A large supply of this paper was made in the beginning of the present year by the noble Sir, my most upright friend, Mr. Alexander Dick, Golden Knight, most worthy President of the Royal Edinburgh College of Physicians. This aquatic plant grows in very great abundance in ponds at his villa near Edinburgh, and I wish you to inform me whether it is *Confera Bullosa* (*Syst. Nat.,* p. 1346, n. 3). He asked me to send to you, with his best wishes, these specimens, along with the paper, named *Linnæa* after his daughter.

Recently I have observed *Ephemera* to shed its skin entirely, and that instar to shed a skin while in the winged stage of its life. I do not know whether this metamorphosis has been observed by others; I have noticed it in all the specimens of *Ephemera.* I have enclosed in this letter a specimen of *Ephemera vulgata* (*Syst. Nat.,* p. 546, n. 1), with shed skin.[5]

---

[4] Bishop Pococke, who had traveled in Asia.

[5] The Ephemerida are the only insects that molt after the winged stage is reached.

I have also sent to you certain specimens of the *Bryum Striatum,* in which, as I hope, you will perceive the clearly visible stamina. During this past summer, with the aid of a barometer and of geometric mensuration, I have begun to note the different heights of the Scottish plants above the surface of the Ocean.

[Notes on Scottish plants]

[Oct. 12, 1762]

LETTER IV

Charles Linnæus sends greeting to Mr. John Walker,
a most distinguished gentleman.

A most friendly letter, written by you last year, in fact, on October 12, I received fourteen days ago; I do not know where it has remained unnoticed so long. With deepest gratitude I acknowledge the esteem with which the illustrious members of the renowned Society of Edinburgh have received me, when they wished to adopt me as a member of their Society. Would that I could return their kindness in a manner worthy of them! You recommended me, I hope without detriment to your reputation.

I am sorry that *Erica Hibernica* has withdrawn its fellow-countrymen. But, if Alpine, I know by experience how difficult it is to find Alpine plants. I am astonished at his skill in making paper from a water-plant; certainly it is a beautiful success; I think that it must be the *Confera Bullata,* everywhere of most frequent occurrence; surely it can be prepared from other water-plants. I certainly wonder that Mr. Alexander Dick could have foretold this. Please greet the noble maid who has inscribed this paper with my name.

In regard to the metamorphosis of the *Ephemeræ,* the fact is very well known among Zoologists. The specimens of *Bryum Striatum* were beautiful, but even now I doubt whether they are true stamens, or another off-shoot: surely in most mosses the *anthera* is concealed in the cover itself, as I have shown in my dissertation on the *Buxbaumia.* I wish that you would undertake the task of proving this, and that you would consult many: you can best do this; I can not, as I am not in the city. Is not the other *Lagopus Plin. Will.* a variety of the ordinary *Lagopus?* I see quite well the marks by which you distinguish it. But the appearance and habits are the same, if the authorities are trustworthy. If I had seen the stuffed skin, I could say for certain.

The specimens of the *Scepter of Charles and Linnæus* I will gladly send. Perchance I can send them to London next spring by D. D. Rothman, inasmuch as I can not send them to England in a letter. It is fine that you have dug up thirty-six different specimens of plants, which I shall carefully note down; but, unless

I myself at least should see the dried specimens, I can not admit them to my collection, since variation so easily deceives the most careful men.

Recently a little work of mine has come from the press; it establishes the species of my plants—a second and larger edition. The sixth volume of *Academic Delights* and the sixth edition of the *Genera of Plants* are now being prepared.

I have not yet been able properly to compare your thirty-five different new species, since I am staying on my country estate during the summer holidays; but as soon as I return to the city and to my library, I shall properly compare them, one and all, and tell you what I observe. The *Lavatera Terna* is certainly my arboreal *Lavatera,* as is quite evident from the calyx which you sent to me. Your *Bryum Striatum* has the same male [△ △ △] in the apex as *Minium dillenii.*

More in the near future.

Upsala, June 20, 1763.

APPENDIX

Walker's Scientific Terminology

The modern character of Walker's scientific vocabulary is impressive. The first use of several geological words and phrases should be credited to him. In other instances, geological terms had been used before but Walker employed them in a more precise manner and related them to definite geological phenomena. He may be credited with using most of them for the first time in a classroom, thus establishing their status and giving them respectability in the new science.

The following list is prepared to show the scope of Walker's geologic knowledge. Names of minerals as such are not included. Unless otherwise designated the word or phrase occurs first in the lecture notes and may be considered to have been in Walker's vocabulary at least as early as 1779. All words without comment were used in the modern sense.

GEOLOGICAL TERMS

ACCIDENTAL STRATA—Included all soils, gravel, and sand deposits, as well as deltaic material. Included what are now termed Recent deposits and most Pleistocene deposits.

ÆRA [ERA]—Usually credited to Hutton (1785) as a geological word.

ARENACEOUS (1765)

ARGILLACEOUS (1758)

ATTRITION

BEDS [of limestone] (1758)

BITUMEN, BITUMINOUS (1765)

BIVALVE SHELLS (1770)

BOWLDER STONES—Under this term Walker described glacial drift.

CALCAREOUS (1758)

CHALK—Substance of animal origin.

COLUMNAR BASALT

CONCRETION—Used in the Hebrides report to refer to any large object in fine matrix, such as a phenocryst or pebble. Observation made in 1764; classified in lecture notes.

CONGLOMERATE—Used in the Hebrides report (1765) as "one in which the ingredients are compacted together without any visible cement." Usually credited to Bakewell, 1813.

CONTRACTED [veins]

CORAL ISLANDS

CORAL PLAINS

CORAL REEFS

DEGRADATION

DELINEATION [of mineral matter] (1772)

DENSITY [of rocks, increased by pressure]

DILATED [veins]

DIP

DISSEMINATED GOLD (1772)

EFFERVESCE, EFFERVESCENT (1758)

ENCROACHMENT [of the sea upon the land]

ENTOMOLOGY—Used as a subject heading in lectures—a very early classroom use of the term.

EPOCH—Usually credited to Playfair, 1802.

EXTRANEOUS FOSSILS (*ca.* 1772)—Fossils in the modern sense.

FERRUGINEOUS MATTER (1764)

FISSILE

FISSURES

FLOCCULENT

FLUCTIVAGOUS—Referring to materials carried by liquids, surface or subsurface; "formed in a solution of water or some other menstruum."

FOSSIL (1770)—Indicating anything dug from the earth, e.g., mineral, rock, or fossil.

FOSSIL PLANTS—Cited as a certain sign of coal.

FOSSILISTS—Geologists in the modern sense.

FOUNTAIN-HEAD—Hydrostatic head.

FRACTURE [in rocks] (1772)

FREESTONE

FRIABLE (*ca.* 1765)

GEODE

GEOLOGY—First use as the subject of a series of lectures and as one of the major divisions of natural history, deserving to stand alone as a separate science.

GEYSER

GEYSERITE—Usually credited to Delametherie, 1812.

GRAIN [of a mineral] (*ca.* 1765)

GRANITE—A rock "composed of three essential ingredients, quartz, mica, and feltspar."

GRANULES (1772)

GRANULOSE, GRANULATED (*ca.* 1765).

GRAVEL (1765)

GRIT, OR GRAIN (1764)—Referring to texture.

GROUND (1765)—Groundmass of a rock.

HANGER SIDE [of vein]—Same as hanging side.

HETEROGENEOUS PARTICLES

HOMOGENOUS (1765)

HORIZONTAL STRATA (1770)

HUMACEOUS

IMBRICATED STRUCTURE (1765)

IMPERVIOUS

INCLINATION, OR DIP

INCRUSTATION—Coating with minerals as a method of fossilization.

INDURATED STRATA (1770)

INDURATION—Compression as a method of fossilization.

INSERTION—The filling of pore spaces with mineral matter as a method of fossilization.

INTERMITTENT SPRINGS

INTERSECTOR—Transverse separation extending from one vein to another.

INTERSTRATUM

LAPIDEOUS (1770)—Rocky.

LONGITUDINAL FRACTURES (1765)

MASSES OF QUARTZ (1764)

MICACEOUS (1765)

OFF-SET—In reference to a mineral vein.

OOLITE

OVER-LAP (1772)—As in the overlap of the Secondary on the Primitive.

PALINGENESIS—Used in describing the growth and development of crystals.

PALPABLE (1765)—Particle visible to the eye.

PARALLEL STRATA

PERIOD—As a geological term usually credited to Lyell, 1833.

PERIODICAL SPRINGS

PERVIOUS—Commonly used by Walker in a geological sense.

PETRIFACTION (1764)

PETRIFIED (1758)

PISOLITHOS—Pisolite.

POROSITY

PORPHYRY (1764)—"A compound rock, consisting of a siliceous ground, with concretions of feltspar."

PRIMARY [mountains]—A synonym for Primitive.

PRIMITIVE MOUNTAINS—Used in referring to those mountains composed of rocks older than Secondary.

PRIMITIVE STRATA (1772)—Unfossiliferous strata below the Secondary; in general equal to Precambrian.

QUARTZOSE (1765)

REGRESS [of the sea from the land]

ROCK CHRONOLOGY—Suggested in early (pre-1779) notes as ascertainable by means of fossils.

SALIENT AND RECEDING ANGLES (1772)—The relationship of mountain valleys and ridges.

SECONDARY MOUNTAINS—Included those mountains composed for the most part of Secondary strata and formed in an "æra" later than the Primary mountains; folding less complex than the Primitive and sediments contain extraneous fossils.

SECONDARY STRATA (1772)—Includes the equivalent of Paleozoic, Mesozoic, and Tertiary.

SILICEOUS (1764)

SLEECH—Mud, mire, or slime in the bed of a river or on the seashore.

SLICKENSIDES

SOLE [of a vein] (1772)

SPECIFIC GRAVITY (1764)

STALACTITES

STALAGMA—Stalagmite.

STRATA [horizontal or inclined] (1758)

STRATH

STRATIFIED GRANITE

STRATULUM (*ca.* 1772)

STRATUM (1758)

STRATUM [of loam intermixed with large stones] (1758)

STRATUM SUPERSTRATUM

STRIAE—Linear grooves on minerals.

STRIATED (1764)—Of stones and minerals having minute grooves.

STRIKE—Usually credited to Sedgwick and Murchison, 1829.

STRUCTURE—Referring to the folding of rocks.

SUBAQUEOUS STRATUM

SUBJACENT STRATUM

SUBMARINE VOLCANOES

SUBSOIL

SUBSTRAMEN—Fine-grained matrix, gluten, or ground, containing larger fragments.

SUPERFICIAL STRATUM (1770)

SUPERINCUMBENT

SUPERINDUCED

SUPERJACENT STRATUM (1720)

SUPERSTRATUM

TERTIARY HILLS—Hills composed for the most part of Accidental strata.

TEXTURE—The characteristics of the particles of rocks and minerals.

TILL-BAND—Applied by Walker to a stiff clay, more or less impervious to water, usually unstratified, often forming a subsoil, and containing pebbles. Originally an agricultural term in Scotland. Walker, having a great interest in agriculture, used the term in 1772 in a geological sense to describe a type of rock.

TRITURATION

TURBINATED SHELLS (1770)

Appendix

UNCTUOUS

VARIEGATED

VEINS (1772)

VENIGENEOUS—Occurring in veins.

VERTICAL STRATA (*ca.* 1765)

VITRIFIED, VITREOUS (1765)

WATER-SHED

MINERALOGICAL TERMS

AMORPHOUS

CLEAVE, CLEAVAGE

COCKS-COMB CRYSTAL
  (1758)

COLUMNAR

CONCHOIDAL FRACTURE

CRYSTAL, CRYSTALLINE,
  CRYSTALLIZE

CUBIC

DODECAHEDRAL

DOUBLE REFRACTION

ELASTICITY

FIBROUS

FLEXIBLE

FOLIATED

FRACTURE (1764)

GRAIN (1764)

HARDNESS

HEXAHEDRAL

IMBRICATED STRUCTURE
  (1764)

IMPALPABLE

LAMELLAE, LAMELLATED
  (1772), LAMELLAR

LAMINATED, LAMINOUS
  (1765)

LINEAR CLEAVAGE

LUSTER

OCTAHEDRAL

OCTANGULAR

OPAQUE

PALPABLE

PELLUCID

PENTAGONIC

PHOSPHORESCENCE

PLATED STRUCTURE

POLYGONAL

POLYHEDRAL

PRISM, PRISMATIC

PYRAMID, PYRAMIDAL

QUADRILATERAL PYRAMIDS

REFLECTION

REFRACT, REFRACTION

RHOMBIC

RHOMBOHEDRAL

RHOMBOIDAL (1765)

SCABROUS

SCINTILLESCENCE

SEMIPELLUCID

STREAK

STRIAE, STRIATED,
  STRIATIONS

TESSULAR

TETRAHEDRA

TRANSLUCENT

TRANSPARENT

TRAPEZIFORM

TRIHEDRA

TRIPEDRAL

VITREOUS

VITRESCENT

BIBLIOGRAPHY

WORKS CITED

Agricola, Georgius. *De natura fossilium.* English translation by Herbert C. Hoover. 1950.

Aldrovandus. *Ulysses Aldrovandi . . . musaeum metallicum* in Libros IV. 1648.

Avicenna. *Kitab al-Shifa (The Book of the Remedy).* English translation by E. J. Holmyard and D. C. Mandeville. Paris, 1927.

Baumer, Johann W. *Historia naturalis regni mineralogici.* Frankfurt, 1780.

Bauschius, Joannes L. *Schediasma.* Jena, 1668.

Becher, Johann Joachim. *Chymisches Laboratorium.* Frankfurt, 1680.

―――. *Natur-Kündigung der Metallen.* Frankfurt, 1679.

―――. *Parnassi illustrati.* Part III: Mineralogia. Ulm, 1663.

―――. *Physica subterraneae.* Frankfurt, 1703.

Bergman, Torbern. *Manuel de minéralogiste ou sciagraphie du règne minéral.* Augmented by J. C. Delamétherie. 2 vols. 1792.

Born, I. von. *Briefe aus Walschland über Natürliche Merkwürdigkeiten.* 1773.

―――. Travels through the Banat of Temeswar, Transylvania, and Hungary. Translated by R. E. Raspe. London, 1777.

Bourguet, Louis. *Traité des pétrifactions.* 1742.

―――. *Du Règne minéral.* Paris. 1771.

Buffon, G. L. *Histoire naturelle des minéraux.* Paris, 1783–88.

Caesalpinus, Andreas. *De metallicis.* Rome, 1596.

―――. *De plantis libri.* Florence, 1583.

Caesius, Bernardus. *Mineralogia, sive naturalis philosophiae thesauri. . . .* Lyons, 1636.

Cardano, Girolamo. *Several commentarii, in Hippocratis de aere, aquis et locis.* 1570.

Cartheuser, Fredrick Augusti. *Elementa mineralogiae.* Frankfurt, 1755.

———. *Mineralogische Abhandlungen.* Giessen, 1771.

Cronstedt, Axel von. *Versuch einer Mineralogie.* Translated from Swedish and enlarged by A. G. Werner. Leipzig, 1780.

Dávila, Pedro Francisco. *Catalogue systématique et raisonné des curiosités de la nature et de l'art qui composent le cabinet de M. Dávila.* . . . Paris, 1767.

De Bomare, M. V. *Minéralogie ou nouvelle exposition du règne minéral.* 1774.

Dioscorides. *Materia medica.*

Donati, Vitaliano. *Essai sur l'histoire naturelle de la Mer Adriatique.* La Haye, 1758.

Dortous de Mairan, Jean Jacques. *Dissertation sur les variations du baromètre.* Bordeaux, 1715.

———. *Traité physique et historique de l'aurore boréale.* Paris, 1733.

Edwards, George. *Natural History.* London, 1776.

Engeström, Gustav von. *An Essay towards a System of Mineralogy.* 1772.

Fabricius, Johann. "Mineralogische und technologische Bemerkungen," *Neue Beyträge zur Mineralgeschichte.* Länder, 1778.

Fallopius, G. *De medicatis aquis atque de fossilibus.* Venice, 1564.

Ferber, Johann. *Travels through Italy in 1771–72.* Translation into English. London, 1776.

Fontana, P. D. Cajetano. *Instituto Physico-Astronomica.* Modena, 1695.

Forster, J. R. *Classification of Fossils and Minerals.* London, 1768.

Forster, Johann G. A. *Observations Made during a Voyage around the World with Cook.* London, 1777.

Fourcroy, Antoine. *Elements of Natural History.* London, 1788.

Gaertner, Josephus. *De fructibus et seminibus plantarum.* . . . Stuttgart, 1788.

Galen. *De simplicium medicamentorum facultatibus.* Paris, 1530.

Gerhard, Carl Abraham. *Versuch einer Geschichte des Mineral Reichs.* Berlin, 1781.

Gesner, Conrad. *De rerum fossilium, lapidum et gemmarum maxime.* . . . Zurich, 1565.

Gmelin, S. G. *Flora sibirica.* 1747.

Grew, Nehemiah. *Musaeum regalis societatis.* London, 1681.

Hales, Stephen. *Statical Essays.* 1738.

Hassenfratz, Jean Henri. *Méthode de nomenclature chimique.* 1787.

Hawkesworth, John. *An Account of the Voyages undertaken by the order of his present Majesty for making discoveries in the Southern Hemisphere.* . . . London, 1773.

Hawkins, Sir Richard. *The Observations of Sir Richard Hawkins in his voyage into the South Sea.* . . . 1593.

Henckel, J. F. *Introduction à la minéralogie.* 2 vols. 1756.

———. *Idée générale de l'origine des pierres, fondée sur les observations et des expériences.* Paris, 1760.

———. *Pyritologie, ou histoire naturelle de la pyrite.* 1725.

Hill, John, M. D. *The History of Fossils.* London, 1748.

Hunter, John. *Essays and observations on Natural History, Anatomy, Physiology, Psychology, and Geology.* Edited by Sir Richard Owen. 1861.

Hutton, James. "Systems of the Earth," *Trans. Roy. Soc. Edinburgh* 1 (1788).

Huygens, Christian. *Traité de la lumière . . . refraction du cristal d'Islande.* 1690.

Justi, J. H. G. *Grundriss des gesammten Mineralreiches.* Göttingen, 1757.

Keir, James. *The first part of a dictionary of chemistry. . . .* Birmingham, 1789.

Kentman, Johann. *Nomenclature rerum fossilium.* 1565.

Kircher, Athanasius. *Mundus subterraneus.* Amsterdam, 1678.

Kirwan, Richard. *Elements of Mineralogy.* London, 1784.

Klaproth, Martin Heinrich. *Observation Relative to the Mineralogical and Chemical History of the Fossils of Cornwall.* London, 1787.

Klein, Jacob. *De lapidibus macrocosmi.* St. Petersburg, 1758.

Lavoisier, Antoine L. *Opuscules physiques et chymiques.* 1774.

Lehmann, Johann Gottlob. *Abhandlung von den Metal-Müttern und der Erzeugung der Metalle.* Berlin, 1753.

———. *Traités de physique d'histoire naturelle, de minéralogie et de métallurgie.* 3 vols. Paris, 1759.

Licetus, Fortunius. *Litheosphorus sive de lapide bononiense lucem in se conceptam ab ambiente claro mox in tenebris mire conservante liber.* 1640.

Ludwig, Christian G. *Terrae musei Dresdensis.* Lippstadt, 1748.

Lunardi, Vincenzo. *An Account of Five Aerial Voyages in Scotland: In a Series of Letters to . . . Chevalier G. Compagni.* London, 1786.

Margraff, F. N. *Diss . . . inaugosistens febrium naturam in genere.* 1749.

Mariotte, Edme. *The Motion of Water and Fluids.* Translated from the French by J. T. Desaguliers. London, 1718.

Maskelyne, Nevil. *An account of observations made on the mountain Schehallien.* London, 1776.

Mendes da Costa, Emanuel. *Natural History of Fossils.* London, 1757.

Moreau de Maupertuis. *Éléments de géographie.* Amsterdam, 1744.

Morton, John. *The Natural History of Northamptonshire; with some account of the antiquities. . . .* London, 1712.

Paracelsus. *Of Metals and Minerals.* London, 1657.

Pitton de Tournefort, Joseph. *Histoire des plantes qui naissent aux environs de Paris, avec leur usage dans la médecine.* Paris, 1698.

Pliny. *Historia naturalis.*

Plot, Robert. *The Natural History of Oxfordshire: Being an Essay towards the Natural History of England.* Oxford, 1677.

Pott, Johann H. *Lithogéognosie ou examen chymique des pierres et des terres.* Paris, 1753.

Price, James. *An Account of Some Experiments on Mercury, Silver, and Gold.* Oxford, 1782.

Ramel, M. R. B. *Consultations de médecine et mémoire sur l'air de Gemenos.* La Haye, 1785.

Raspe, R. E. *See* Born, I. von.

Ray, John. *Observations, topographical, moral, and physiological.* . . . London, 1673.

Razoumovsky, Count Grigory. *Voyages minéralogiques.* Lausanne, 1784.

Romé de Lisle, Jean Baptiste Louis. *Cristallographie.* 4 vols. 1783.

———. *Des caractères extérieurs de minéraux.* 1784.

———. *Essai de cristallographie.* 4 vols. Paris, 1772.

Sage, B. G. *Examen chymique.* 1769.

———. *Élémens de minéralogie docimastique.* Paris, 1772.

Saunders, William. *Elements of the Practice of the Physic.* . . . London, 1780.

Saussure, Horace Benedict de. *Voyages dans les Alpes.* 4 vols. Neuchâtel, 1779–96.

Scheele, Carl W. *Mémoires de Chymie.* Dijon, 1785.

Scheuchzer, Johann Jacob. *Sciagraphia lithologica.* Danzig, 1740.

Scopoli, Giovanni Antonio. *Principia mineralogiae systematicae et practicae.* Prague, 1772.

Solinus, Caius Julius. *Collectanea rerum memorabilium.* Translated by T. Mommsen. Berlin, 1864.

Starrman, Andres. *A Voyage to the Cape of Good Hope.* London, 1785.

Theophrastus. *Concerning Stones.* Translated by John Hill. London, 1746.

Volta, Ab. *Elementi di mineralogia.* Pavia, 1787.

Wallerius, Johann G. *Mineralogie.* Paris, 1753.

———. *Systema mineralogicum.* 2 vols. Stockholm, 1778.

Whitehurst, John. *An Inquiry into the Original State and Formation of the Earth.* London, 1778.

Woltersdorf, J. L. *Systema minerale.* Berlin, 1738.

Woodward, John. *Fossils of all Kinds.* London, 1728.

———. *An Essay toward a Natural History of the Earth.* 1695.

Works by Walker

"An Account of a New Medicinal Well, Lately Discovered Near Moffat," *Phil. Trans. Roy. Soc. London,* 50, Part 1 (1758): 117–47. (Read Feb. 10 and March 3, 1757.)

"An Account of the Irruption of Solway Moss in Dec. 16, 1772," *ibid.,* 62 (1772): 123–27. (Read Feb. 13, 1772.)

Schediasma Fossilium. Edinburgh, 1781.

"On the Flowers of Muscous Plants," Natural History Society 1 (1782): 203–10 (manuscript).

"Description of a Whale," *ibid.,* pp. 89–98 (manuscript).

"The Rise of Sap in Trees," *Trans. Roy. Soc. Edinburgh* 1 (1788): 3–41. (Read Dec. 8, 1783, and Jan. 5, 1785.)

"On the Subterranean Heat," Natural History Society 8 (1789): 175–86 (manuscript).

"A Sermon Preached before His Majestey's High Commissioners, May 19, 1791" (28 pp., copy in National Library of Scotland).

"An Essay on Peat, Containing an Account of Its Origin, of its Chymical Principles and General Properties," *Prize Essays and Trans. of the Highland Soc. of Scotland* 2 (1803): 1–137.

"Extracts from an Essay on the Natural, Commercial, and Economical History of the Herring," *ibid.,* 2 (1803): 270–305.

"On the Natural History of the Samon," *ibid.,* pp. 346–76.

"On the Cattle and Corn of the Highlands," *ibid.,* pp. 164–204.

"An Essay on Kelp," *ibid.,* 1 (1799): 1–31. (Read to the Society in 1788.)

An Economical History of the Hebrides and Highlands of Scotland. 2 vols. Edinburgh, 1808.

Essays on Natural History and Rural Economy. Edinburgh, 1808. Reprinted London, 1812. (Published by his friend Charles Stewart.)

All "Essays" written between 1764 and 1774 except 1, 7–9, 14–15. Dates of manuscripts are given in parentheses. The manuscripts are in the collection of the Library of the University of Edinburgh.

1. A Catalogue of some of the most considerable trees in Scotland, pp. 1–82; Fruit trees, pp. 82–90 (*ca.* 1796).
2. Natural History of the inhabitants of the Highlands, pp. 91–110.
3. History of the Island of Icolumbkil [Iona], pp. 111–217; Fossils, minerals and rocks, plants, and animals, pp. 152–217.
4. History of the Island of Jura, pp. 219–83 (*ca.* 1765); Plants and Fossils, pp. 242–83.
5. A description of the Basse and its Production, pp. 283–313; Animal and Plants, pp. 289–313.
6. The History of Shell Marle, pp. 313–23 (*ca.* 1770).
7. Public Lectures, Anno 1788, on the Utility and Progress of Natural History and Manner of Philosophising, pp. 323–47 (1788).
8. Memoirs of Sir Andrew Balfour, pp. 347–69 (post-1782).
9. The Natural History of Lock-Leven, pp. 371–84 (*ca.* 1797).
10. Mineralogical Journal from Edinburgh to Elliock, pp. 385–95 (*ca.* 1772).
11. Mineralogical Journal from Edinburgh to London, pp. 395–402 (probably October, 1765).
12. Salicetum; The botanical history and cultivation of Willows, pp. 403–70.
13. Mammalia Scotica, pp. 471–536 (1797).
14. Statistical Account of the Parish of Collington, pp. 535–615; Plants and Animals, pp. 588–604 (1795).

15. Memorial Concerning the Scarcity of Grain in Scotland, pp. 615–29 (December, 1800).

"Notice of Mineralogical Journeys, and of a Mineralogical System, by the late Rev. Dr. John Walker," *Edinburgh Phil. J.* 6 (1822): 88–95.

MANUSCRIPTS

[N.B. Except as otherwise noted, manuscripts are in the Edinburgh University Library.]

An Essay on the cultivation of sand with Marle. D.C. 2-392.

Novum organum botanicum. Outlines a series of suggestions for a complete system of botany under the auspices of the Royal Society of Edinburgh. D.C. 2-392.

Annotations on Professor Walker's *Schediasma Fossilium 1781.* D.C. 8-20.

Essays, transcripts and other papers, containing "A statistical account of the parish of Colington." . . . 3 vols. D.C. 1-57–59.

Extracts . . . relating to the North of Scotland, . . . 3 vols. D.C. 2-36–38.

Lectures on natural history. 1782. D.C. 2-22.

Doctor Walker's Lectures on Natural History. Edinburgh, May 12, 1784. Notes by Thomas Charles Hope. 5 vols. Hope Bequest.

Lectures on natural history. 1790. 6 vols., in part zoological. D.C. 2-23–28.

Notes from Dr. W.'s lectures on natural history. 1791. Taken by John Douglas. D.C. 8-31.

Dr. Walker's Lectures. July 1791. . . . 4 vols. D.C. 7-113, 1–3.

Notes of Professor J. Walker's lectures on Natural History. Dr. S. W. Carruthers, 1933. D.C. 10-33.

Lectures on natural history. 5 vols., in part zoological. D.C. 2-17–21.

Letter to C. Stewart, printer, with instructions for printing his manuscript. Colinton. La. III. 353. *See also* letter of C. Stewart to J. Ferrier about works of J. Walker to be printed posthumously.

WORKS CONTAINING REFERENCES TO WALKER

A natural history of the island of Icolumbkil. La. III. 575.

Lectures in Agriculture. Aberdeen University Library (microfilm University of Illinois).

Notes and memoranda on natural history [mostly botanical]. 1766–75. 7 vols. D.C. 2-29–35.

Occasional remarks, essays, etc. 137 pages. D.C. 2-40.

Papers and memorials by Professor John Walker and many letters to him from friends and fellow naturalists, including letters from Linnaeus, Brugmans, Fabricius, Pennant, etc. La. III. 352.

Papers on agriculture, natural history, etc. 2 vols. D.C. 2-39, 1–2.

Papers on natural history, etc. D.C. 1-18.

Sermons. 1758–90. Includes sermons preached at Glencorse, Moffat, Colington, and to the General Assembly when Moderator, 1790. Holograph. La. III. 132.

Arnot, Hugo. *The History of Edinburgh*. Edinburgh 1816. Pp. 310–12.

Bower, A. *The History of the University*. 3 vols. Edinburgh, 1817–30. III, 218, 228.

Catalogue of books, manuscripts . . . in the British Museum. London, 1940. VIII, Supplement, 26, 1986.

Catalogue of printed books in the Library of the University of Edinburgh. Edinburgh, 1923. III, 1109.

Dictionary of National Biography. London, 1909. XX, 531.

Finlayson, Charles P. *Records of Scientific and Medical Societies Preserved in the University Library*. Edinburgh, 1958. I, No. 3, 14–19.

Fleming, John. Review of Hooker's *Flora scotica, New Edinburgh Review* 1 (1821): 467.

———. *A History of British Animals*. 1828. P. 550.

———. *The Lithology of Edinburgh*. 1859.

Grant, Alexander. *History of the University of Edinburgh*. 1884.

Greville, R. K. *Algae britannicae*. 1830. Vol. III.

Guion, Fred. J. "Poem to the Rev. Mr. Walker," *Scots Mag.* 34 (1772): 372, 441.

Headrick. *View of the mineralogy, agriculture, manufactures and fisheries of the Island of Arran, with notice of antiquities*. 1807.

———. *General view of the Agriculture of the County of Angus, or Fifeshire*. 1813.

Jameson, Laurence. "Biographical memoir of the late Professor Jameson in Edinburgh," *New Phil. J.* 57 (1854): 7.

Jameson, R. *Outline of the Mineralogy of the Scottish Isles*. Edinburgh, 1800.

———. *Mineralogy of the Shetland Islands*. Edinburgh, 1798.

———. *A System of Mineralogy*. Edinburgh, 1820.

Jardine, Sir W. "Memoir of John Walker, D.D." in *The Birds of Great Britain and Ireland*. 1842. III, 3–50.

Johnston, George. *Flora of Berwick-upon-Tweed*. 1829.

———. *The Botany of the Eastern Borders*. 1859.

Kay, John. *A Series of Original Portraits and Caricature Etchings*. Edinburgh, 1837–40. III, 178.

———. *A Descriptive Catalogue of Original Portraits*. Edinburgh, 1836.

Loudon, J. C. *Arboretum et fruticetum britannicum*.

Macvicar, Symers. *Ann. Scot. Nat. Hist.* (1895), p. 257.

Murray, Thomas. *Annals of the Parish of Colington*. Edinburgh, 1863. Pp. 64–68.

Ritchie, James. "Natural History and the Emergence of Geology in the Scottish Universities," *Trans. Edinburgh Geol. Soc.* 15 (1952): 297–316.

———. "A Double Centenary—Two Notable Naturalists, Robert Jameson and Edward Forbes," *Proc. Roy. Soc. Edinburgh, B,* 66, Pt. 1 (1956): 29–58.

The Royal Scottish Museum. 1954. P. 36.

Bibliography

Scott, Hew. "The Succession of Ministers in the Church of Scotland," *Fasti ecclesiae scoticanae*. Edinburgh, 1915. I, 4.

Shankie, David. *The Parish of Colington*. Edinburgh, 1902. Pp. 54–55.

Smallwood, W. M. Extracts from two unpublished letters from Linnaeus, *Sci. Monthly* 49 (1939): 65–70.

Smith, Lady Pleasance. *Memoir and Correspondence of the late Sir James Edward Smith, M.D.* London, 1832.

Taylor, George. "John Walker, D.D., F.R.S.E., 1731–1803, A Notable Scottish Naturalist," *Trans. Bot. Soc. Edinburgh* 38 (1959): 180–203.

Turnbull, W. Robertson. *History of Moffat*. 1871.

Tytler, Alexander Fraser [Lord Woodhouselee]. *Memoirs of the Life and Writings of Henry Home of Kames*. 1807. Vols. 1 and 2.

INDEX

Index